IN ASSOCIATION WITH

SQA

HODDER GIBSON

Model Papers
WITH ANSWERS

PLUS: Official SQA Specimen Paper & 2015 Past Paper With Answers

Higher for CfE
Geography

2014 Specimen Question Paper, Model Papers & 2015 Exam

HODDER GIBSON
AN HACHETTE UK COMPANY

This book contains the official 2014 SQA Specimen Question Paper and 2015 Exam for Higher for CfE Geography, with associated SQA approved answers modified from the official marking instructions that accompany the paper.

In addition the book contains model papers, together with answers, plus study skills advice. These papers, some of which may include a limited number of previously published SQA questions, have been specially commissioned by Hodder Gibson, and have been written by experienced senior teachers and examiners in line with the new Higher for CfE syllabus and assessment outlines, Spring 2014. This is not SQA material but has been devised to provide further practice for Higher for CfE examinations in 2014 and beyond.

Hodder Gibson is grateful to the copyright holders, as credited on the final page of the Answer Section, for permission to use their material. Every effort has been made to trace the copyright holders and to obtain their permission for the use of copyright material. Hodder Gibson will be happy to receive information allowing us to rectify any error or omission in future editions.

Hachette UK's policy is to use papers that are natural, renewable and recyclable products and made from wood grown in sustainable forests. The logging and manufacturing processes are expected to conform to the environmental regulations of the country of origin.

Orders: please contact Bookpoint Ltd, 130 Park Drive, Milton Park, Abingdon, Oxon OX14 4SE. Telephone: (44) 01235 827720. Fax: (44) 01235 400454. Lines are open 9.00–5.00, Monday to Saturday, with a 24-hour message answering service. Visit our website at www.hoddereducation.co.uk. Hodder Gibson can be contacted direct on: Tel: 0141 848 1609; Fax: 0141 889 6315; email: hoddergibson@hodder.co.uk

This collection first published in 2015 by
Hodder Gibson, an imprint of Hodder Education,
An Hachette UK Company
2a Christie Street
Paisley PA1 1NB

Typeset by Aptara, Inc.

Printed in the UK

A catalogue record for this title is available from the British Library

ISBN: 978-1-4718-6075-1

3 2 1

2016 2015

Introduction

Study Skills – what you need to know to pass exams!

Pause for thought

Many students might skip quickly through a page like this. After all, we all know how to revise. Do you really though?

Think about this:

"IF YOU ALWAYS DO WHAT YOU ALWAYS DO, YOU WILL ALWAYS GET WHAT YOU HAVE ALWAYS GOT."

Do you like the grades you get? Do you want to do better? If you get full marks in your assessment, then that's great! Change nothing! This section is just to help you get that little bit better than you already are.

There are two main parts to the advice on offer here. The first part highlights fairly obvious things but which are also very important. The second part makes suggestions about revision that you might not have thought about but which WILL help you.

Part 1

DOH! It's so obvious but …

Start revising in good time

Don't leave it until the last minute – this will make you panic.

Make a revision timetable that sets out work time AND play time.

Sleep and eat!

Obvious really, and very helpful. Avoid arguments or stressful things too – even games that wind you up. You need to be fit, awake and focused!

Know your place!

Make sure you know exactly **WHEN and WHERE** your exams are.

Know your enemy!

Make sure you know what to expect in the exam.

How is the paper structured?

How much time is there for each question?

What types of question are involved?

Which topics seem to come up time and time again?

Which topics are your strongest and which are your weakest?

Are all topics compulsory or are there choices?

Learn by DOING!

There is no substitute for past papers and practice papers – they are simply essential! Tackling this collection of papers and answers is exactly the right thing to be doing as your exams approach.

Part 2

People learn in different ways. Some like low light, some bright. Some like early morning, some like evening/night. Some prefer warm, some prefer cold. But everyone uses their BRAIN and the brain works when it is active. Passive learning – sitting gazing at notes – is the most INEFFICIENT way to learn anything. Below you will find tips and ideas for making your revision more effective and maybe even more enjoyable. What follows gets your brain active, and active learning works!

Activity 1 – Stop and review

Step 1

When you have done no more than 5 minutes of revision reading STOP!

Step 2

Write a heading in your own words which sums up the topic you have been revising.

Step 3

Write a summary of what you have revised in no more than two sentences. Don't fool yourself by saying, "I know it, but I cannot put it into words". That just means you don't know it well enough. If you cannot write your summary, revise that section again, knowing that you must write a summary at the end of it. Many of you will have notebooks full of blue/black ink writing. Many of the pages will not be especially attractive or memorable so try to liven them up a bit with colour as you are reviewing and rewriting. **This is a great memory aid, and memory is the most important thing.**

Activity 2 – Use technology!

Why should everything be written down? Have you thought about 'mental' maps, diagrams, cartoons and colour to help you learn? And rather than write down notes, why not record your revision material?

What about having a text message revision session with friends? Keep in touch with them to find out how and what they are revising and share ideas and questions.

Why not make a video diary where you tell the camera what you are doing, what you think you have learned and what you still have to do? No one has to see or hear it, but the process of having to organise your thoughts in a formal way to explain something is a very important learning practice.

Be sure to make use of electronic files. You could begin to summarise your class notes. Your typing might be slow, but it will get faster and the typed notes will be easier to read than the scribbles in your class notes. Try to add different fonts and colours to make your work stand out. You can easily Google relevant pictures, cartoons and diagrams which you can copy and paste to make your work more attractive and **MEMORABLE**.

Activity 3 – This is it. Do this and you will know lots!

Step 1

In this task you must be very honest with yourself! Find the SQA syllabus for your subject (www.sqa.org.uk). Look at how it is broken down into main topics called MANDATORY knowledge. That means stuff you MUST know.

Step 2

BEFORE you do ANY revision on this topic, write a list of everything that you already know about the subject. It might be quite a long list but you only need to write it once. It shows you all the information that is already in your long-term memory so you know what parts you do not need to revise!

Step 3

Pick a chapter or section from your book or revision notes. Choose a fairly large section or a whole chapter to get the most out of this activity.

With a buddy, use Skype, Facetime, Twitter or any other communication you have, to play the game 'If this is the answer, what is the question?'. For example, if you are revising Geography and the answer you provide is "meander", your buddy would have to make up a question like "What is the word that describes a feature of a river where it flows slowly and bends often from side to side?".

Make up 10 "answers" based on the content of the chapter or section you are using. Give this to your buddy to solve while you solve theirs.

Step 4

Construct a wordsearch of at least 10 × 10 squares. You can make it as big as you like but keep it realistic. Work together with a group of friends. Many apps allow you to make wordsearch puzzles online. The words and phrases can go in any direction and phrases can be split. Your puzzle must only contain facts linked to the topic you are revising. Your task is to find 10 bits of information to hide in your puzzle, but you must not repeat information that you used in Step 3. DO NOT show where the words are. Fill up empty squares with random letters. Remember to keep a note of where your answers are hidden but do not show your friends. When you have a complete puzzle, exchange it with a friend to solve each other's puzzle.

Step 5

Now make up 10 questions (not "answers" this time) based on the same chapter used in the previous two tasks. Again, you must find NEW information that you have not yet used. Now it's getting hard to find that new information! Again, give your questions to a friend to answer.

Step 6

As you have been doing the puzzles, your brain has been actively searching for new information. Now write a NEW LIST that contains only the new information you have discovered when doing the puzzles. Your new list is the one to look at repeatedly for short bursts over the next few days. Try to remember more and more of it without looking at it. After a few days, you should be able to add words from your second list to your first list as you increase the information in your long-term memory.

FINALLY! Be inspired...

Make a list of different revision ideas and beside each one write **THINGS I HAVE** tried, **THINGS I WILL** try and **THINGS I MIGHT** try. Don't be scared of trying something new.

And remember – "FAIL TO PREPARE AND PREPARE TO FAIL!"

Higher Geography

The Course

To gain the course award, candidates must pass the learning outcomes for all three units as well as the course assessment. The purpose of the course assessment is to assess added value of the course, confirm attainment and provide a grade.

The Exam

The course assessment is structured into Component 1, which is an externally set and marked question paper worth 60 marks, and Component 2, which is an assignment based on a research topic chosen by the candidate worth 30 marks. 90 marks are available in total.

Component 1 – The External Question Paper

The question paper is set and marked by SQA, and conducted in centres under exam conditions. Candidates are expected to complete this question paper in 2 hours and 15 minutes. The questions will be asked on a local, regional and global scale and the paper has four sections.

- **Section 1: Physical Environments – 15 marks**
  Candidates must answer all questions in this section.

- **Section 2: Human Environments –15 marks**
  Candidates must answer all questions in this section.

- **Section 3: Global Issues – 20 marks**
  Candidates must answer **two** questions from out of **five** options. Candidates must answer all parts of the question within each chosen option.

- **Section 4: A question on the application of Geographical Skills – 10 marks**
  This section consists of an extended response question requiring the learner to apply geographical skills acquired during the course. These may include interpreting an Ordnance Survey map, using six-figure grid references, understanding/using scale, direction and distance. You will also be required to extract and interpret information from a variety of sources. Candidates must answer this question.

The question paper component of the course assessment will have **a greater emphasis on the assessment of knowledge and understanding** than the assignment. The other marks will be awarded for the demonstration of skills.

Component 2 – The Assignment

The purpose of the assignment is to show challenge and application by demonstrating skills, knowledge and understanding within the context of a geographical topic or issue. Candidates can choose the topic or issue to be researched.

This assignment provides candidates with the opportunity to demonstrate higher-order cognitive skills and knowledge of methods and techniques.

- The assignment involves identifying a geographical topic or issue.

- It also involves carrying out research, which should include fieldwork where appropriate.

- Candidates will be asked to demonstrate knowledge of the suitability of the methods and/or reliability of the sources used.

- The assignment will involve drawing on detailed knowledge and understanding of the topic or issue.

- It will also involve analysing and processing information from a range of sources.

- It will require reaching a conclusion supported by a range of evidence on a geographical topic or issue.

- Candidates should demonstrate the skill of communicating information.

The assignment is set by centres within SQA guidelines, and the assessment is conducted under a high level of supervision and control by the presenting centre. The production of evidence for assessment will be conducted within 1 hour and 30 minutes and with the use of specified resources. This evidence is then submitted to SQA for external marking.

The assignment component of the course assessment will have **a greater emphasis on the assessment of skills** than the question paper.

Layout of this Book

This book contains three model papers, which mirror the actual SQA exam as much as possible. The layout, paper colour and question level are all similar to the actual exam that you will sit, so that you are familiar with what the exam paper will look like.

The answer section is at the back of the book. Each answer contains a worked-out answer or solution so that you can see how to arrive at the right answer. The answers also include practical tips on how to tackle certain types of questions, details of how marks are awarded and advice on just what the examiners will be looking for.

As well as your class notes and textbooks, these model papers are a useful revision tool because they will help you to get used to answering exam-style questions. You may find as you work through the questions that they refer to a case study or an example that you haven't come across before. Don't worry! You should be able to transfer your knowledge of a topic or theme to a new example. The enhanced answer section at the back will demonstrate how to read and interpret the question to identify the topic being examined and how to apply your course knowledge in order to answer the question successfully.

Examination Hints

- Make sure that you have read the instructions in the question carefully and that you have avoided needless errors such as answering the wrong sections, failing to explain when asked to or perhaps omitting to refer to a named area or case study.

- If you are asked for a named country or city, make sure you include details of any case study you have covered.

- Avoid vague answers when asked for detail. For example, avoid vague terms such as "dry soils" or "fertile soils". Instead, try to provide more detailed information in your answer such as "deep and well-drained soils" or "rich in nutrients".

- If you are given data in the form of maps, diagrams and tables in the question, make sure you refer to this information in your answer to support any points of view that you give.

- Be guided by the number of marks for a question as to the length of your answer.
- Make sure that you leave yourself sufficient time to answer all of the questions.
- One technique which you might find helpful, especially when answering long questions, is to "brainstorm" possible points for your answer. You can write these down in a list at the start of your answer and cross them out as you go through them.
- If you have any time left in the exam, use it to go back over your answers to see if you can add anything to what you have written by way of additional text or including more examples or diagrams which you may have omitted.

Common Errors

Lack of sufficient detail

- This often occurs in Higher case study answers, especially in questions with high marks.
- Many candidates fail to provide sufficient detail in answers, often by omitting reference to specific examples, or by failing to elaborate or develop points made in their answer.
- Remember that you have to give more information in your answers to gain a mark.

Irrelevant answers

- You must read the question instructions carefully to avoid giving answers which are irrelevant.
- For example, if asked to explain and you simply describe you will not score marks. If asked for a named example and you do not provide one, you will forfeit marks.

Statement reversals

- Occasionally, questions involve opposites. For example, some answers would say "death rates are high in developing countries due to poor health care" and then go on to say "death rates are low in developed countries due to good health care". Avoid doing this. You are simply stating the reverse of the first statement.
- A better second statement might be that "high standards of hygiene, health and education in developed countries have helped to bring about low death rates".

Repetition

- You should be careful not to repeat points already made in your answer. These will not gain any further marks. You may feel that you have written a long answer, but it could contain the same basic information repeated over and over. Unfortunately, these statements will be recognised and ignored when your paper is marked.

Listing

- If you give a simple list of points rather than fuller statements in your answer, you may lose marks, for example, in a 5-mark question you will obtain only 1 mark for a list.

Bullet points

- The same rule applies to a simple list of bullet points. However, if you give bullet points with some detailed explanation, you could achieve full marks.

Types of Questions and Command Words

In these model papers, and in the exam itself, a number of command words will be used in the different types of questions you will encounter. The command words are used to show you how you should answer a question; some words indicate that you should write more than others. If you familiarise yourself with these command words, it will help you to structure your answers more effectively. The question types to look out for are listed below.

Explain

These questions ask you to explain and give reasons, for example, "strategies" and relationships. If resources are provided in the question, make sure you refer to them in your answer. Some marks may be allowed for description but these will be quite restricted.

Analyse

This involves identifying parts and the relationships between them by showing the links between different components and related concepts, noting similarities and differences, explaining possible consequences and implications, and explaining the impact of, for example, processes of degradation, strategies adopted to control events and government policies on people and the environment.

Evaluate

This will involve making judgements on, for example, the relative success or failure of strategies and projects such as a river basin management scheme or aid programmes.

Discuss

These questions ask you to develop your thoughts, for example, on a specific project or change in specified situations. You may be asked to consider both sides of an argument and provide a range of comments on each viewpoint.

Geographical Skills

This question is designed to examine your geographical skills. You will be given an Ordnance Survey map along with a variety of other resources which might include a climate graph, information table, newspaper article, statistics and photographs. You will be given a set of conditions to follow; these should be used to help you answer the question. You should use/evaluate the information you have been given to make judgements and back up your answer with map evidence, information from the diagrams and tables, etc. Remember the diagrams and OS map are there for a purpose and contain valuable information that you should incorporate in your answer!

Good luck!

Remember that the rewards for passing Higher Geography are well worth it! Your pass will help you get the future you want for yourself. In the exam, be confident in your own ability. Watch your time and pace yourself carefully. If you're not sure how to answer a question, trust your instincts and just give it a go anyway – keep calm and don't panic! GOOD LUCK!

2014 Specimen Question Paper

National Qualifications
SPECIMEN ONLY

SQ20/H/01   **Geography**

Date — Not applicable

Duration — 2 hours and 15 minutes

Total marks — 60

SECTION 1 — PHYSICAL ENVIRONMENTS — 15 marks

Attempt ALL questions.

SECTION 2 — HUMAN ENVIRONMENTS — 15 marks

Attempt ALL questions.

SECTION 3 — GLOBAL ISSUES — 20 marks

Attempt TWO questions.

SECTION 4 — APPLICATION OF GEOGRAPHICAL SKILLS — 10 marks

Attempt the question.

Credit will be given for appropriately labelled sketch maps and diagrams.

Write your answers clearly in the answer booklet provided. In the answer booklet you must clearly identify the question number you are attempting.

Use **blue** or **black** ink.

Before leaving the examination room you must give your answer booklet to the Invigilator; if you do not you may lose all the marks for this paper.

MARKS

SECTION 1: PHYSICAL ENVIRONMENTS — 15 marks

Attempt ALL questions

Question 1

> Corries are landscape features present in glaciated upland areas.

Explain the conditions and processes involved in the formation of a corrie.

You may wish to use an annotated diagram or diagrams. **5**

Question 2

Look at Diagram Q2.

Explain how factors such as those shown in the diagram affect the formation of a brown earth soil. **6**

Diagram Q2: Main factors affecting soil formation

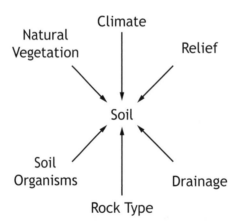

Question 3

Explain why there is a surplus of solar energy in the tropical latitudes and a deficit of solar energy towards the poles.

You may wish to use an annotated diagram or diagrams. **4**

SECTION 2: HUMAN ENVIRONMENTS — 15 marks

Attempt ALL questions

Question 1

> Nigeria conducted a population census in 2006. However, the chairperson of the National Population Commission stated in 2012 that 'Nigeria has no data. People can't really tell you precisely what the population is'. Another census will be conducted in 2016.

Explain the problems of collecting accurate population data in developing countries.

6

Question 2

> 2012 saw a significant increase in Germany's population. This was not due to a sudden baby boom, but to the many immigrants moving to the country. Experts point out this could result in both benefits and problems.

Referring to a named case study, analyse the impact of migration on **either** the donor country **or** the receiving country.

5

Question 3

Referring to **either** a named rainforest **or** a named semi-arid area, explain the techniques used to combat rural land degradation.

4

SECTION 3: GLOBAL ISSUES — 20 marks

Attempt TWO questions

MARKS

Question 1 — River Basin Management

(a)   Study Map/Data Q7 and Table Q7

Explain why there is a need for water management in Ghana.   5

Map/Data Q7

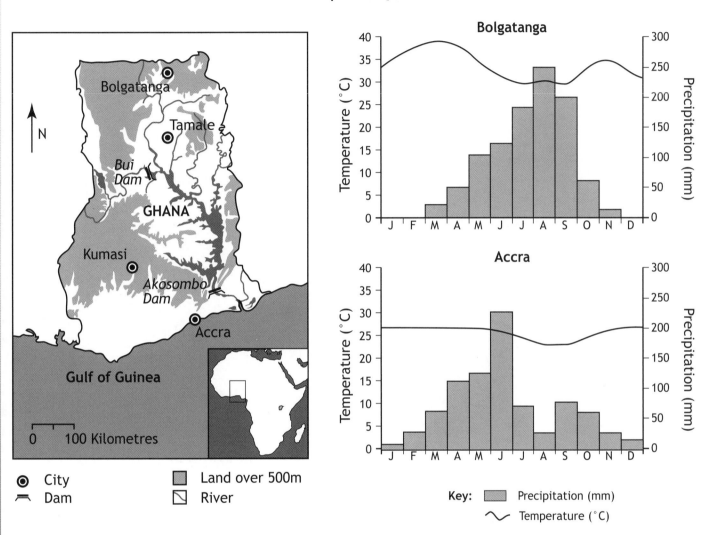

Table Q7: Ghana Development Data

Current population	24·28 million
Projected population by 2040	35·8 million
Labour force by occupation	56% in agriculture 15% in industry 29% in services
% of population with access to electricity	45%

(b)   Explain the negative impacts of any named water management project.

In your answer you must refer to socio-economic **and** environmental impacts.   5

MARKS

Question 2 — Development and Health

For malaria or any other water-related disease that you have studied:

(a)   explain the methods used to try and control the spread of the disease; **and**

(b)   evaluate the effectiveness of these methods.   10

MARKS

Question 3 — Global Climate Change

Look at Diagrams Q9a and Q9b.

(a)   Explain the human activities which have contributed to the changes in global air temperatures.   5

Diagram Q9a: Global air temperatures 1850—2011

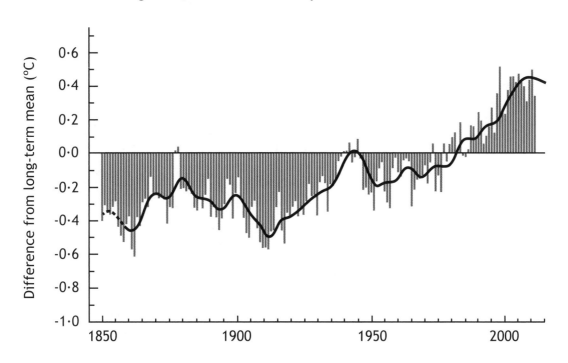

(b)   Discuss the possible impacts of global warming throughout the world.   5

Diagram Q9b: Greenhouse gases, emissions by type

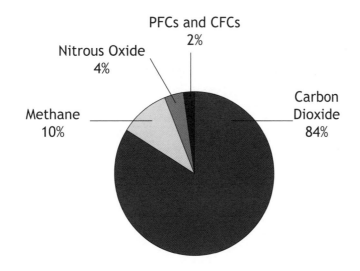

MARKS

Question 4 — Trade, Aid and Geopolitics

(a) Study Table Q10.

Suggest reasons for the inequalities in trade shown in the table below.   5

Table Q10: Selected development indicators

		Germany	Thailand	Kenya
Trade statistics (million US$, 2012)	Exports	1,492,000	226,200	5,942
	Imports	1,276,000	217,800	14,390
	Balance of trade	216,000	8,400	−8,448
Economic Indicators (2012)	GDP per capita (US$) (PPP*)	39,100	1,800	10,300
	% employed in agriculture	2%	38%	75%
	% employed in manufacturing	24%	14%	10%
	% employed in services	74%	48%	5%
	Main exports	Motor vehicles, machinery, chemicals, computer and electronic products	Electronics, computer parts, automobiles and parts, electrical appliances, machinery and equipment, textiles	Tea, horticultural products, coffee, petroleum products, fish

PPP* = purchasing power parity

(b) Explain the strategies used to reduce inequalities in world trade.   5

MARKS

Question 5 — Energy

(a)   Look at Graph Q11.

Explain the differences in energy consumption between developed and developing countries.

4

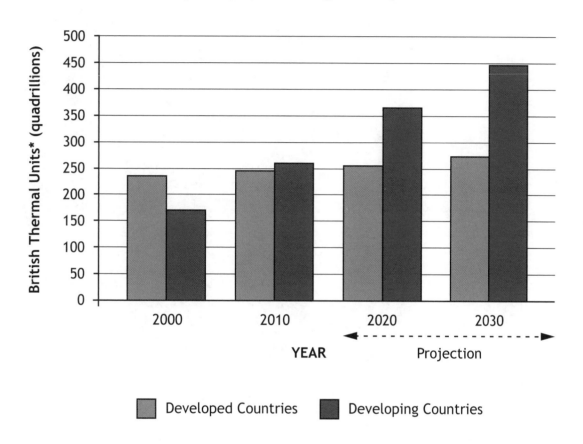

Graph Q11: World energy consumption

(b)   Referring to different countries, evaluate the suitability of renewable approaches to generating energy.

6

OS Map (Extract 1349/EXP272: Lincoln)

Extract No 1349/EXP272

1:25 000 Scale
Explorer Series

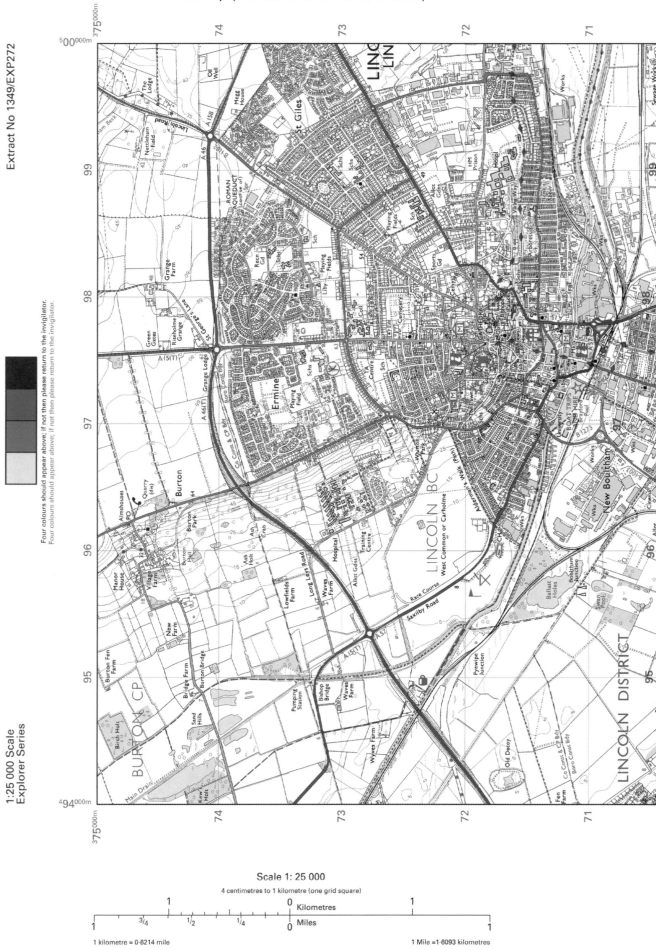

Scale 1: 25 000

4 centimetres to 1 kilometre (one grid square)

Kilometres

Miles

1 kilometre = 0·6214 mile   1 Mile = 1·6093 kilometres

SECTION 4: APPLICATION OF GEOGRAPHICAL SKILLS — 10 marks

Attempt the question

Question 1

The city of Lincoln has decided to hold a 10 k race. Working to the brief below, a route has been proposed.

Brief for Lincoln 10 k race

The route should:

- be suitable for all participants

- cause minimum disruption to people and business in the local area

- promote business in the local area

- have a suitable start/finish line

- be scenic/interesting for participants.

Study Map Q12: Proposed 10 k Route; OS Map (Extract 1349/EXP272: Lincoln); Diagram Q12; and Graph Q12.

Referring to map evidence and other information from the sources, evaluate the suitability of the proposed route (Map Q12) in relation to the brief for the 10 k race.

You should suggest possible improvements to the route.

10

Question 1 (continued)

Map Q12: Proposed 10 k route

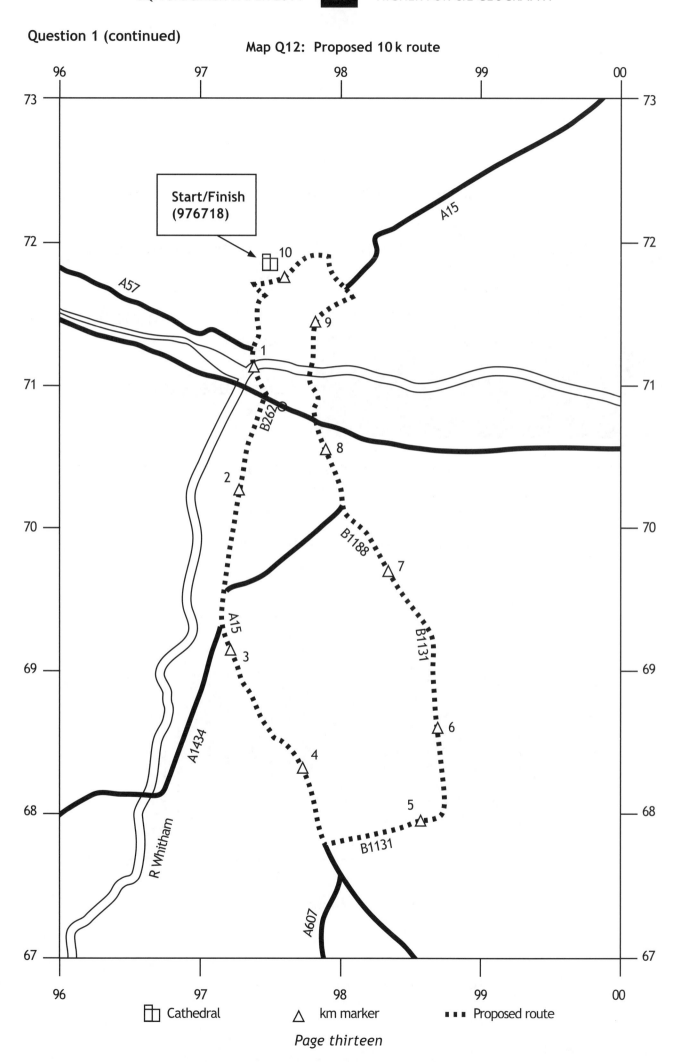

⌂ Cathedral   △ km marker   ▪▪▪ Proposed route

Question 1 (continued)

Diagram Q12

Lincoln 10 k Run
Sunday 16th February

Join over 5,000 people taking part

Live music from local bands

Safe route along closed roads

For further information on the race and nearby accommodation go to:
www.visitlincoln.co.uk

Graph Q12: Visitor numbers to Lincoln

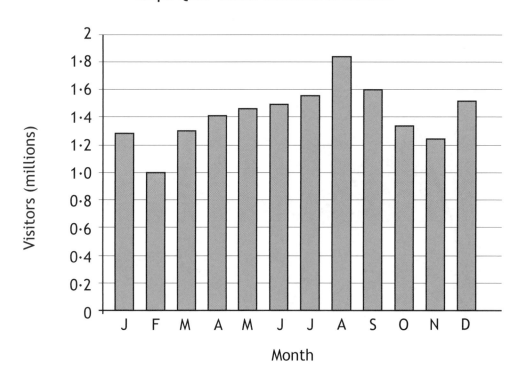

[END OF SPECIMEN QUESTION PAPER]

HIGHER FOR CfE

Model Paper 1

Whilst this Model Paper has been specially commissioned by Hodder Gibson for use as practice for the Higher (for Curriculum for Excellence) exams, the key reference documents remain the SQA Specimen Paper 2014 and SQA Past Paper 2015.

 HODDER GIBSON
LEARN MORE

**National
Qualifications
MODEL PAPER 1**

Geography

Duration — 2 hours and 15 minutes

Total marks — 60

SECTION 1 — PHYSICAL ENVIRONMENTS — 15 marks

Attempt ALL questions.

SECTION 2 — HUMAN ENVIRONMENTS — 15 marks

Attempt ALL questions.

SECTION 3 — GLOBAL ISSUES — 20 marks

Attempt TWO questions.

SECTION 4 — APPLICATION OF GEOGRAPHICAL SKILLS — 10 marks

Attempt the question.

Credit will be given for appropriately labelled sketch maps and diagrams.

Write your answers clearly in the answer booklet provided. In the answer booklet you must clearly identify the question number you are attempting.

Use **blue** or **black** ink.

Before leaving the examination room you must give your answer booklet to the Invigilator; if you do not you may lose all the marks for this paper.

**HODDER
GIBSON**
LEARN MORE

MARKS

SECTION 1: PHYSICAL ENVIRONMENTS — 15 marks

Attempt ALL questions

Question 1

Explain why there is a surplus of solar energy in the tropical latitudes and a deficit of solar energy towards the poles. You may wish to use an annotated diagram or diagrams in your answer.   5

Question 2

Look at Diagram Q2 which shows a coastal area in the UK.

Explain, with the aid of a diagram or diagrams, how feature A was formed.   5

Diagram Q2: Pembrokeshire coast

Question 3

Explain ways in which human activity can affect the hydrological cycle.   5

MARKS

SECTION 2: HUMAN ENVIRONMENTS — 15 marks

Attempt ALL questions

Question 1

Explain the problems of conducting a census in a developing country such as Nigeria.

5

Question 2

Referring to either a rainforest or semi-arid area you have studied, explain the consequences of land degradation on the people and environment.

6

Question 3

For any named developed world city you have studied, explain methods which have been introduced to manage traffic in the Central Business District.

4

SECTION 3: GLOBAL ISSUES — 20 marks

Attempt TWO questions

MARKS

Question 1 — River Basin Management

Study Diagrams Q1A, Q1B, Q1C and Q1D.

(a) Explain why there is a need for water management in the Missouri River Basin.   5

Diagram Q1A:
USA spring flood risk 2012

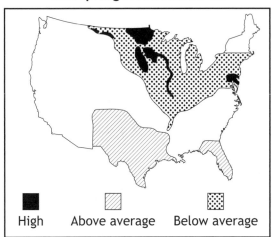

High   Above average   Below average

Diagram Q1B:
USA July temperatures 2012

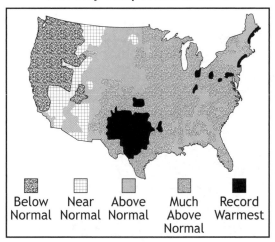

Below Normal   Near Normal   Above Normal   Much Above Normal   Record Warmest

Diagram Q1C: Missouri River Basin

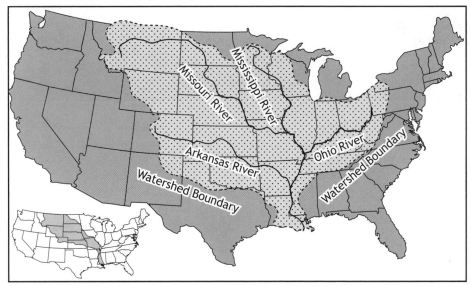

Table Q1D: Population Upper Missouri River Basin

	2010	2013	% Change
Metropolitan Areas			
Great Falls	81,327	82,384	+1.3
Micropolitan Areas			
Bozeman	89,513	94,720	+5.8
Helena	74,801	76,850	+2.7
TOTAL	164,314	171,570	+4.4
Rest of Basin	65,542	64,053	+0.8
Montana	989,415	1,015,165	+2.6

Source: U.S. Census Bureau, Population Division

(b) For any named water management scheme you have studied, explain the positive impacts on the people and environment of the area.   5

MARKS

Question 2 — Development and Health

Look at Diagram Q2.

Explain how primary health care strategies can improve the health and development of a developing country.

10

Diagram Q2: Mobile health clinic, South Africa

The mobile clinic is equipped with medical essentials.

MARKS

Question 3 — Global Climate Change

Look at Diagram Q3.

   (a) Discuss the human causes of global warming.   **5**

   (b) Explain the global impacts of global warming.   **5**

Diagram Q3: Changes in Arctic summer ice 1979–2012

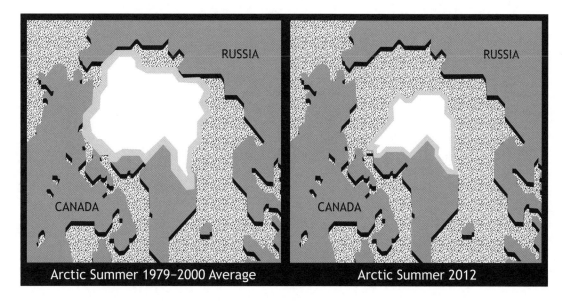

MARKS

Question 4 — Trade, Aid and Geopolitics

Explain why aid from developed countries is often more beneficial to the donor country than the receiving country.

10

MARKS

Question 5 — Energy

> "Unlike fossil fuels, the wind, the Sun and the Earth itself provide fuel that is free, in amounts that are effectively limitless."
>
> Quote – AL GORE

Look at the quote above. With reference to named countries, discuss the suitability of generating electricity from renewable resources.

10

1:50 000 Scale
Landranger Series

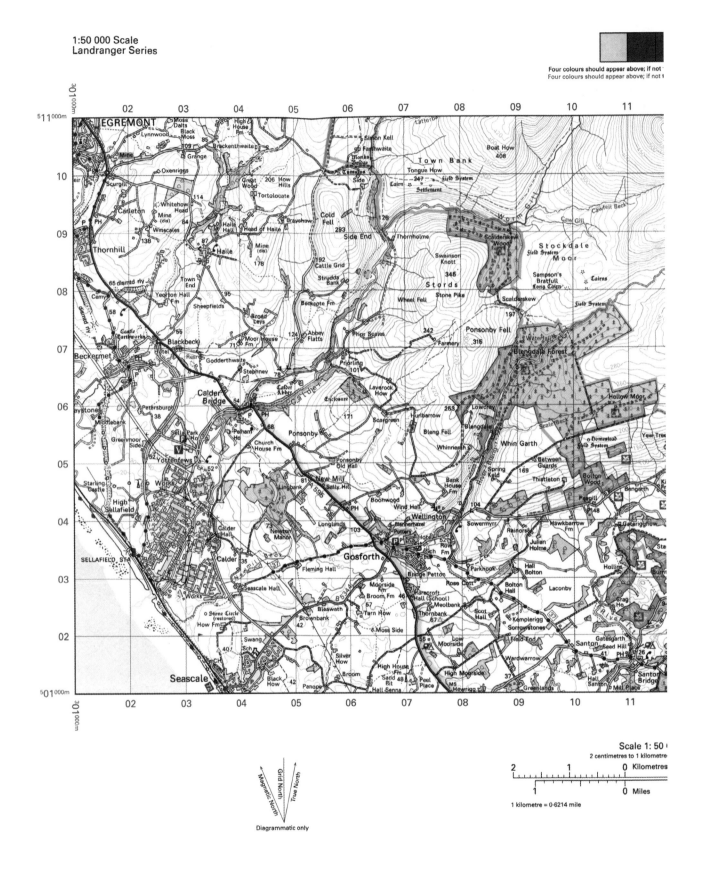

Grid North
True North
Magnetic North

Diagrammatic only

Scale 1: 50
2 centimetres to 1 kilometre

2   1   0 Kilometres

1   0 Miles

1 kilometre = 0·6214 mile

Extract No 1787/89

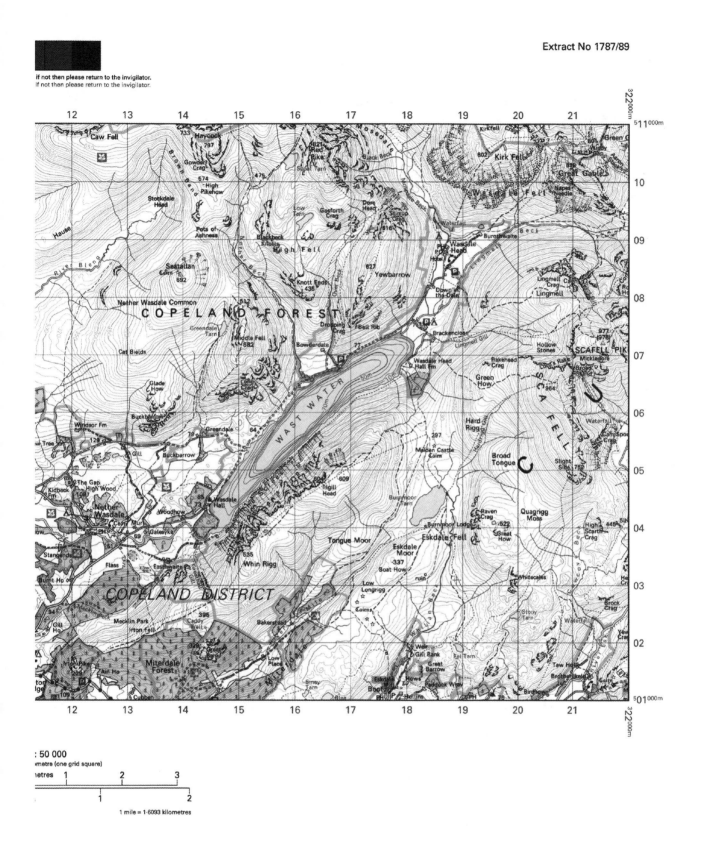

: 50 000

metre (one grid square)

1 mile = 1·6093 kilometres

MARKS

Question 1

It has been proposed to build a new nuclear facility close to the decommissioned* nuclear plant at Sellafield (*no longer produces nuclear power). Working to the brief below, a 618 acre site, directly north of the current plant, on the outskirts of the Lake District National Park, has been proposed.

Brief for new nuclear facility

The site should:
- have minimal impact on the landscape, environment, people and wildlife
- flat land to build on
- have good transportation routes
- a workforce nearby.

Study the O.S. Map Extract: 1787/89 and Diagrams Q1–Q5.

Referring to map evidence and other information from the sources, **evaluate the suitability of the proposed site** in relation to the brief for the new nuclear facility close to Sellafield.

10

Diagram Q1: Proposed site of new nuclear facility

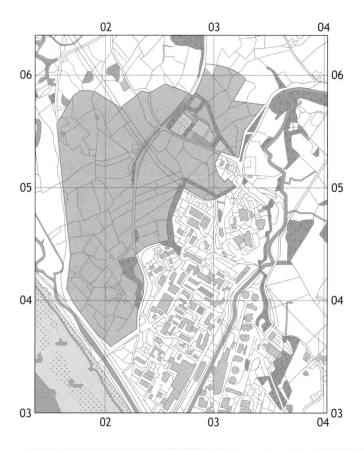

MARKS

Diagram Q2: View of proposed site
(Taken from Cold Fell (058092) — within the Lake District National Park)

Diagram Q3: Sellafield wins 8 safety awards

Sellafield Ltd has won eight awards in the prestigious RoSPA Occupational Health and Safety Awards 2015. The company has recently recorded one of its best ever safety performances. "It makes me very proud to be able to say that despite being one of the most complex nuclear sites in the world, we are also one of the safest, and that is not by chance. As one of the largest industrial sites in Europe with some very unique challenges, our attention to detail and relentless focus on delivering our mission safely means that we have an excellent safety record." Head of Safety for the company, Pete Oldfield said.

Diagram Q4: View of anti-nuclear campaigner

Anti-nuclear campaigner Janine Allis-Smith's son was diagnosed with leukaemia in 1983. She is convinced he was exposed to radiation during family trips to the Cumbrian seaside. Janine says: "He put handfuls of mud and sand on his head and face. I'm sure Sellafield has something to do with it. I know lots of children who've died and whose fathers worked at Sellafield. The graveyard at a church near Newbiggin has lots of graves of children who died in the sixties, seventies and eighties. It was not just leukaemia, but other cancers. Some were stillborn, while other suffered unexplained deaths at a very young age."

MARKS

Diagram Q5: Front Page of Local Newspaper

THE
Lake District News

ALL ABOUT THE BIG WORLD WE LIVE IN　　　　　*EXCLUSIVE NEWS TODAY*

BEACHES NEARS SELLAFIELD CONTAMINATED WITH OVER 1,200 RADIOACTIVE HOTSPOTS

"Dangerous Particles May Remain Undetected"

A record number of radioactive hotspots have been found contaminating public beaches near the Sellafield nuclear complex in Cumbria, according to a report by the site's operator.

As many as 383 radioactive particles and stones were detected and removed from seven beaches in 2010–11, bringing the total retrieved since 2006 to 1,233.

Although Sellafield insists that the health risks for beach users are "very low", there are concerns that some potentially dangerous particles may remain undetected and that contamination keeps being found.

"Beaches Should Be Closed"

Anti-nuclear campaigners have called for beaches to be closed, or for signs to be erected warning the public of the pollution. But the government's Health Protection Agency (HPA) has said "no special precautionary actions are required at this time to limit access to, or use of, beaches."

[END OF MODEL PAPER]

Model Paper 2

Whilst this Model Paper has been specially commissioned by Hodder Gibson for use as practice for the Higher (for Curriculum for Excellence) exams, the key reference documents remain the SQA Specimen Paper 2014 and SQA Past Paper 2015.

For reasons of space and copyright, SECTION 4 has been omitted from this Model Paper.

HODDER GIBSON
LEARN MORE

National
Qualifications
MODEL PAPER 2

Geography

Duration — 2 hours and 15 minutes

Total marks — 60

SECTION 1 — PHYSICAL ENVIRONMENTS — 15 marks

Attempt ALL questions.

SECTION 2 — HUMAN ENVIRONMENTS — 15 marks

Attempt ALL questions.

SECTION 3 — GLOBAL ISSUES — 20 marks

Attempt TWO questions.

Credit will be given for appropriately labelled sketch maps and diagrams.

Write your answers clearly in the answer booklet provided. In the answer booklet you must clearly identify the question number you are attempting.

Use **blue** or **black** ink.

Before leaving the examination room you must give your answer booklet to the Invigilator; if you do not you may lose all the marks for this paper.

MARKS

SECTION 1: PHYSICAL ENVIRONMENTS — 15 marks

Attempt ALL questions

Question 1

Look at Diagram Q1.

Choose two of the factors shown on Diagram Q1. Explain how your chosen factors affect a storm hydrograph.

4

Diagram Q1: Factors affecting a storm hydrograph

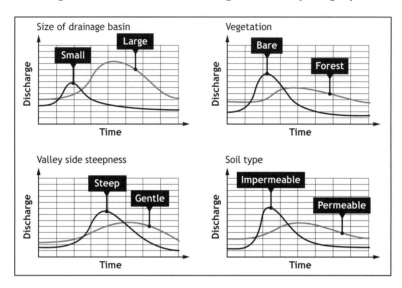

Question 2

Study Diagram Q2.

Choose one of the soil profiles from Diagram Q2. Explain how your chosen soil was formed.

6

Diagram Q2: Selected soil profiles

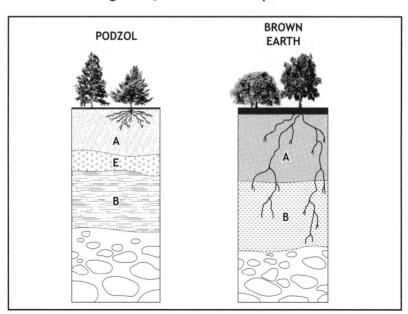

Question 3

With the aid of annotated diagrams, explain the formation of a corrie.

5

MARKS

SECTION 2: HUMAN ENVIRONMENTS — 15 marks

Attempt ALL questions

Question 1

Look at Diagram Q1.

With reference to a migration flow you have studied, explain the impact on the receiving country.

5

Diagram Q1: Recent migrations in Europe

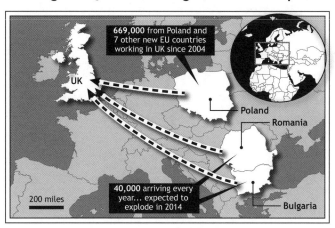

669,000 from Poland and 7 other new EU countries working in UK since 2004

UK

Poland

Romania

200 miles

40,000 arriving every year... expected to explode in 2014

Bulgaria

Question 2

Look at Table Q2.

Referring to either a rainforest or semi-arid area you have studied, explain the strategies used to reduce rural land degradation.

6

Table Q2: Some causes of rural land degradation

North America	Africa north of the Equator OR The Amazon Basin	
Monoculture	Deforestation	Deforestation
Deep ploughing	Overcultivation	Cattle ranching
Farming marginal land	Overgrazing	Mining
Demand for wheat	Population increase	HEP schemes

Question 3

Look at Diagram Q3.

For a named area you have studied, evaluate strategies used to improve shanty towns.

4

Diagram Q3: Changes in shanty towns

Before Upgrading   After Upgrading

Page three

SECTION 3: GLOBAL ISSUES — 20 marks

Attempt TWO questions

MARKS

Question 1 — River Basin Management

(a) Discuss the physical factors that should be taken into consideration when choosing the site of a dam.   **5**

(b) Evaluate the social, economic and environmental advantages of a named water control project you have studied.   **5**

MARKS

Question 2 — Development and Health

For malaria, cholera or bilharzia:

    (a) explain efforts used to combat the disease; and

    (b) evaluate how successful these measures have been.   **10**

MARKS

Question 3 — Global Climate Change

Look at Diagram Q3.

   (a) Discuss ways in which people can reduce/manage the impact of global warming.

6

   (b) Evaluate the effectiveness of these attempts.

4

Diagram Q3: Effects of global warming

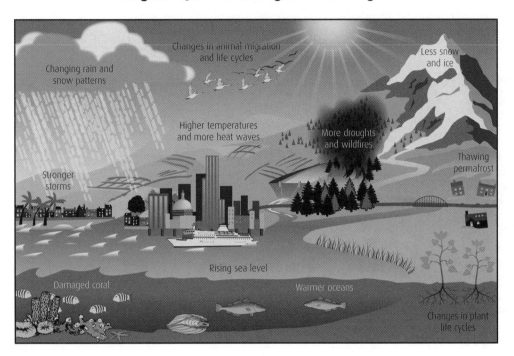

MARKS

Question 4 — Trade, Aid and Geopolitics

Study Diagram Q4.

(a) Explain why the pattern of world trade shown in Diagram Q10 mainly benefits developed countries.

5

(b) Discuss strategies which can be used to reduce inequalities in world trade.

5

Diagram Q4: World trade

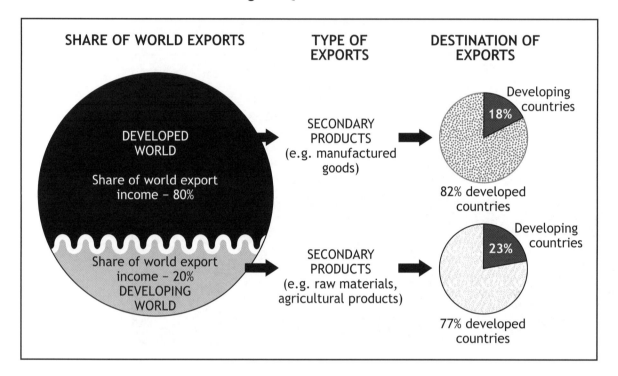

MARKS

Question 5 — Energy

Look at Diagram Q5.

Explain the reasons for the increase in energy consumption for both developed and developing areas of the world.

10

Diagram Q5: Growing world energy demand

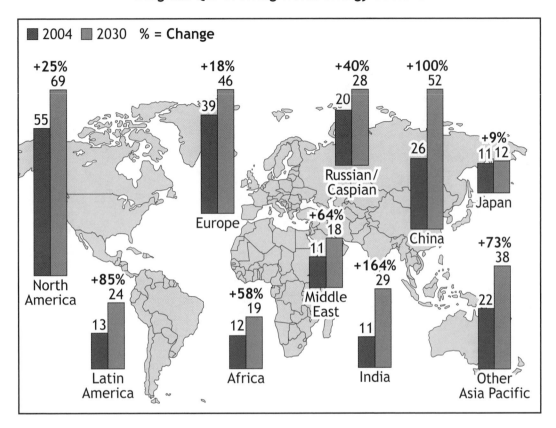

[END OF MODEL PAPER]

HIGHER FOR CfE

Model Paper 3

Whilst this Model Paper has been specially commissioned by Hodder Gibson for use as practice for the Higher (for Curriculum for Excellence) exams, the key reference documents remain the SQA Specimen Paper 2014 and SQA Past Paper 2015.

For reasons of space and copyright, SECTION 4 has been omitted from this Model Paper.

National
Qualifications
MODEL PAPER 3

Geography

Duration — 2 hours and 15 minutes

Total marks — 60

SECTION 1 — PHYSICAL ENVIRONMENTS — 15 marks

Attempt ALL questions.

SECTION 2 — HUMAN ENVIRONMENTS — 15 marks

Attempt ALL questions.

SECTION 3 — GLOBAL ISSUES — 20 marks

Attempt TWO questions.

Credit will be given for appropriately labelled sketch maps and diagrams.

Write your answers clearly in the answer booklet provided. In the answer booklet you must clearly identify the question number you are attempting.

Use **blue** or **black** ink.

Before leaving the examination room you must give your answer booklet to the Invigilator; if you do not you may lose all the marks for this paper.

MARKS

SECTION 1: PHYSICAL ENVIRONMENTS — 15 marks

Attempt ALL questions

Question 1

Look at Diagrams Q1A and Q1C, and Map Q1B.

Explain the variation in rainfall within West Africa.

6

Diagram Q1A: Location of selected air masses and the ITCZ in January and July

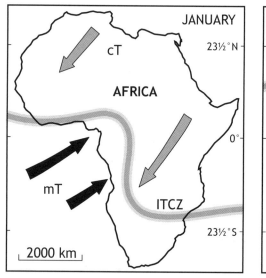
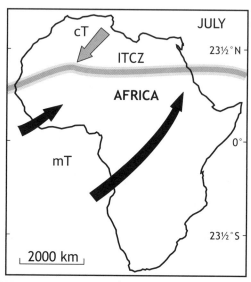

Key:
mT = Maritime Tropical   cT = Continental Tropical
ITCZ = Inter Tropical Convergence Zone

Q1B: Map of West Africa

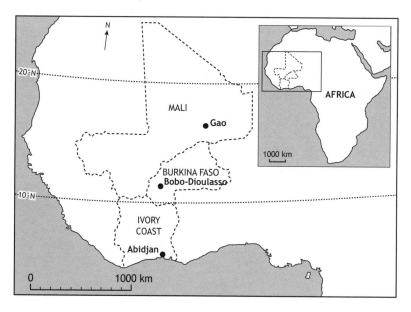

MARKS

Diagram Q1C: Average monthly rainfall/days with precipitation

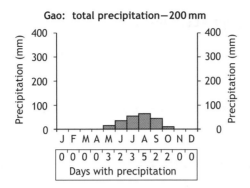

Gao: total precipitation—200 mm

J	F	M	A	M	J	J	A	S	O	N	D
0	0	0	0	3	2	3	5	2	2	0	0

Days with precipitation

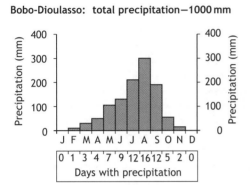

Bobo-Dioulasso: total precipitation—1000 mm

J	F	M	A	M	J	J	A	S	O	N	D
0	1	3	4	7	9	12	16	12	5	2	0

Days with precipitation

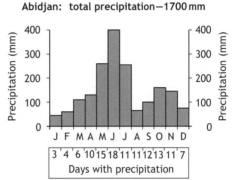

Abidjan: total precipitation—1700 mm

J	F	M	A	M	J	J	A	S	O	N	D
3	4	6	10	15	18	11	11	12	13	11	7

Days with precipitation

Question 2

Explain, with the aid of a diagram or diagrams, how one of the following features of glacial deposition is formed:

- terminal moraine
- esker
- drumlin

5

Question 3

Look at Diagram Q3.

Choose two factors from Diagram Q3 and explain how these factors can influence the formation of brown earth soil.

4

Diagram Q3: Some factors affecting soil formation

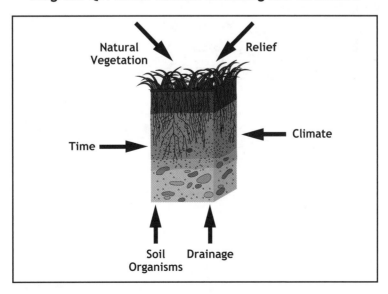

Page three

SECTION 2: HUMAN ENVIRONMENTS — 15 marks

Attempt ALL questions

Question 1

Look at Diagram Q1.

Discuss the possible consequences of the projected 2050 population structure of India for the future economy and the welfare of its citizens.

5

Diagram Q1: Population structure 2010–2050 (projected)

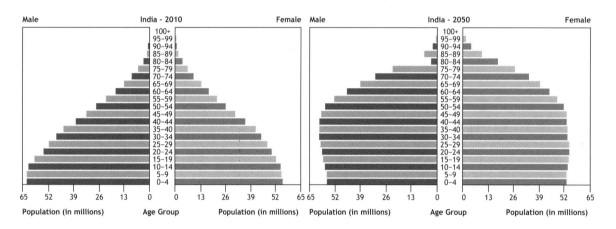

Question 2

For a semi-arid area you have studied, explain the effects of rural land degradation on the people and environment.

5

Question 3

Look at Diagram Q3.

For an area you have studied, evaluate the impact of schemes to improve shanty towns.

5

Diagram Q3: Improving shanty towns

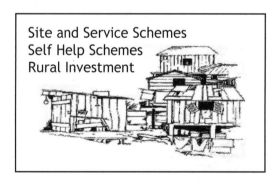

Site and Service Schemes
Self Help Schemes
Rural Investment

SECTION 3: GLOBAL ISSUES — 20 marks

Attempt TWO questions

MARKS

Question 1 — River Basin Management

(a) Look at Diagram Q1A.

Evaluate the suitability of either site A, B or C for the location of a dam.   5

Diagram Q1A: Possible sites of a dam

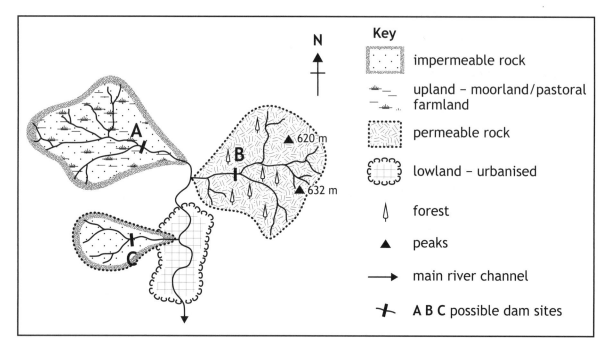

(b) Look at Diagram Q1B

For any named river basin scheme you have studied, discuss the social and environmental disadvantages of your chosen scheme.   5

Diagram Q1B: Narmada River Scheme, India

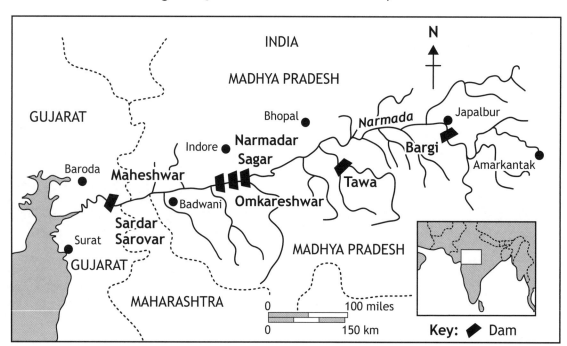

MARKS

Question 2 — Development and Health

Composite measures such as PQLI (Physical Quality of Life Indicator) or HDI (Human Development Indicator) can be used to compare the levels of development between countries.

For one of these composite indicators, or any other composite indicator you have studied:

(a) explain the ways it measures the level of development of a country; and

(b) evaluate its usefulness.   **10**

MARKS

Question 3 — Global Climate Change

Look at Diagram Q3.

(a) Explain the human activities which have contributed to the effects of global warming.   6

(b) Discuss the effects of global warming on people and the environment.   4

Diagram Q3: Global temperature change 1881–2010

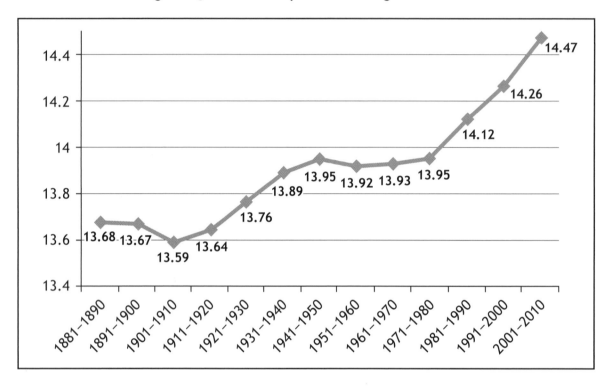

Question 4 — Trade, Aid and Geopolitics

Explain ways in which geopolitics influences foreign aid programmes. You should refer to named examples in your answer.

10

MARKS

Question 5 — Energy

Study Diagram Q5.

Explain the distribution of global energy consumption.

10

Diagram Q5: Distribution of global energy consumption 2014

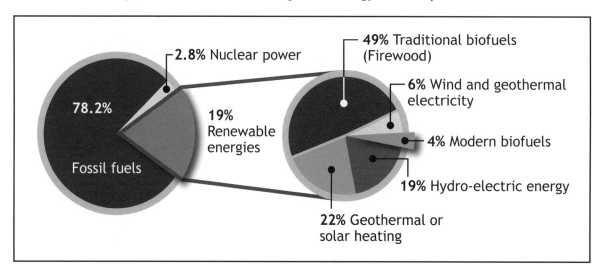

2.8% Nuclear power

49% Traditional biofuels (Firewood)

6% Wind and geothermal electricity

78.2%

19% Renewable energies

4% Modern biofuels

Fossil fuels

19% Hydro-electric energy

22% Geothermal or solar heating

[END OF MODEL PAPER]

HIGHER FOR CfE

2015

National Qualifications 2015

X733/76/11   **Geography**

THURSDAY, 21 MAY

9:00 AM – 11:15 AM

Total marks — 60

SECTION 1 — PHYSICAL ENVIRONMENTS — 15 marks

Attempt ALL questions.

SECTION 2 — HUMAN ENVIRONMENTS — 15 marks

Attempt ALL questions.

SECTION 3 — GLOBAL ISSUES — 20 marks

Attempt TWO questions.

SECTION 4 — APPLICATION OF GEOGRAPHICAL SKILLS — 10 marks

Attempt the question.

Credit will be given for appropriately labelled sketch maps and diagrams.

Write your answers clearly in the answer booklet provided. In the answer booklet you must clearly identify the question number you are attempting.

Use **blue** or **black** ink.

Before leaving the examination room you must give your answer booklet to the Invigilator; if you do not you may lose all the marks for this paper.

MARKS

SECTION 1: PHYSICAL ENVIRONMENTS – 15 marks
Attempt ALL questions

Question 1

Study Diagram Q1 before answering this question.

Diagram Q1: Flood Hydrograph for the River Valency at Boscastle, 16 August 2004

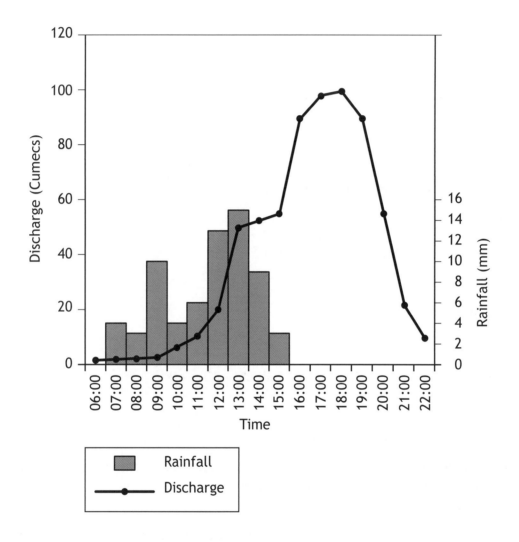

Explain the changes in discharge level of the River Valency at Boscastle on 16 August 2004.   4

MARKS

Question 2

Explain, with the aid of annotated diagrams, the various stages and processes involved in the formation of:

(a)   a stack; **and**

(b)   a sand spit.   7

[Turn over

MARKS

Question 3

Look at Diagram Q3 before answering this question.

Diagram Q3: Surface winds and pressure zones

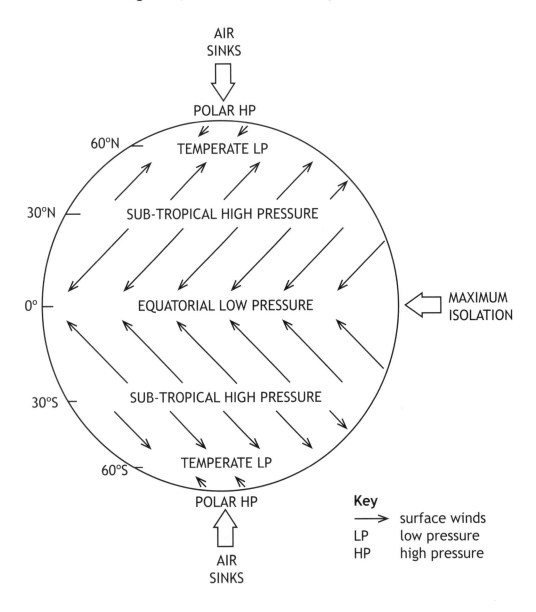

AIR
SINKS

POLAR HP

60°N   TEMPERATE LP

30°N   SUB-TROPICAL HIGH PRESSURE

0°   EQUATORIAL LOW PRESSURE   MAXIMUM
ISOLATION

30°S   SUB-TROPICAL HIGH PRESSURE

60°S   TEMPERATE LP

POLAR HP   **Key**
→ surface winds
LP   low pressure
HP   high pressure

AIR
SINKS

Explain how atmospheric circulation cells and the associated surface winds assist in redistributing energy around the world.

4

SECTION 2: HUMAN ENVIRONMENTS – 15 marks

Attempt ALL questions

MARKS

Question 4

Study Diagrams Q4A and Q4B before answering this question.

Diagram Q4A: Population Pyramid for Ghana, 2013

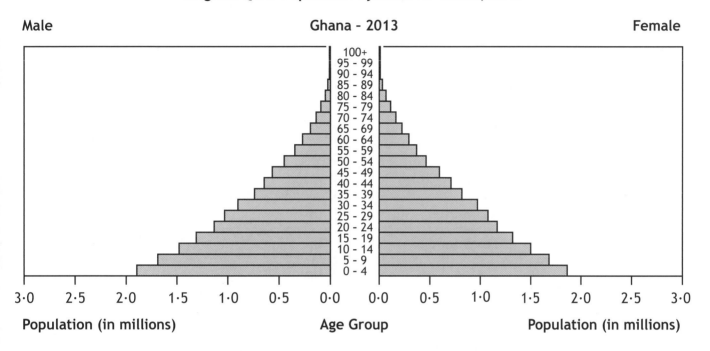

Diagram Q4B: Population Pyramid for Ghana, 2050 (predicted)

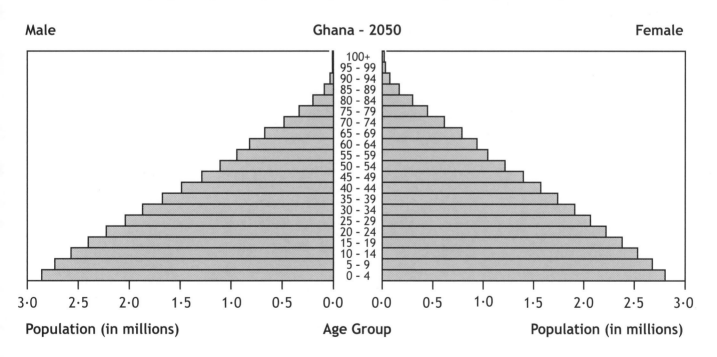

Discuss the possible consequences for Ghana of the 2050 population structure.

5

MARKS

Question 5

Look at Diagram Q5 before answering this question.

Traffic congestion is a major problem in cities in the UK and across the **developed** world.

Diagram Q5: Most congested UK cities, 2013

Rank	City	Congestion trend
1	Belfast	Up
2	Bristol	Level
3	Brighton	Up
4	Edinburgh	Down
5	London	Up
6	Leeds/Bradford	Down
7	Manchester	Up
8	Leicester	Up
9	Sheffield	Up
10	Liverpool	Up

Explain the strategies employed to combat the problems of traffic congestion in a **developed** world city you have studied. You should refer to specific named examples from your chosen city.

5

MARKS

Question 6

Look at Diagram Q6 before answering this question.

Rapid urbanisation in **developing** world cities has resulted in many housing problems.

Diagram Q6: Photograph of Dharavi Slum, Mumbai, India

gary yim / Shutterstock.com

Evaluate the impact of strategies employed to manage housing problems in a **developing** world city you have studied.

5

[Turn over

SECTION 3: GLOBAL ISSUES – 20 marks
Attempt TWO questions

MARKS

Question 7: River Basin Management

(a)   **Explain** the human **and** physical factors which need to be considered when selecting a site for a major dam and its associated reservoir.   5

(b)   Referring to a water control project you have studied, **explain** the **positive** social **and** economic impacts created by the construction of a major dam and its associated reservoir.   5

[Turn over

Question 8: Development & Health

Diagram Q8: Development indicators for selected developing countries

Country	Indicators of Development					
	GDP (US $ per capita)	Employment agriculture (%)	Adult literacy (%)	Birth rate (births per 1000)	Life expectancy at birth (in years)	Hospital beds (per 1000)
Mexico	15,400	14	93	18	76	1·7
Brazil	11,700	16	90	15	73	2·3
Cuba	10,200	20	99	10	78	5·1
Kenya	1,800	75	87	30	63	1·4
Malawi	800	90	75	40	53	1·3

(a) Study Diagram Q8 above before answering this question.

In what ways does the information in the table suggest that the five countries are at different levels of development? **4**

(b) **Suggest reasons** for the wide variations in development which exist between developing countries. You may wish to refer to countries that you have studied. **6**

MARKS

Question 9: Global Climate Change

Look at Diagram Q9.

Diagram Q9: Natural and Enhanced Greenhouse Effect

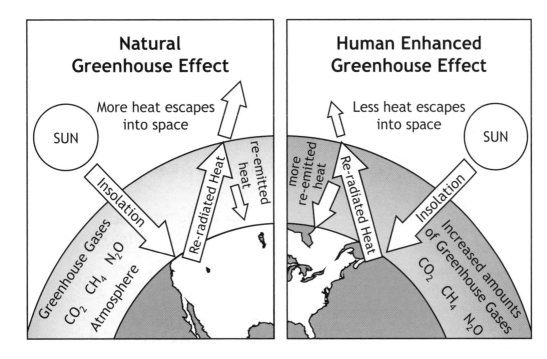

Many scientists believe that human activity has led to an enhanced greenhouse effect.

(a)   **Explain** the human factors that may lead to climate change.   4

(b)   **Discuss** a range of possible effects of climate change. You should support your answer with specific examples.   6

[Turn over

MARKS

Question 10: Trade, Aid and Geopolitics

Read the quotation in Diagram Q10a below.

Diagram Q10a: Quotation from Nelson Mandela (22nd November 2000)

> "Where globalisation means, as it so often does, that the rich and powerful now have new means to further enrich and empower themselves at the cost of the poorer and weaker, we have a responsibility to protest in the name of universal freedom."

(a) **Explain** the social and economic impacts of unfair trade on people and countries in the **Developing World**.

4

Diagram Q10b: Fair Trade and Normal Coffee prices

(b) Study Diagram Q10b.

Using the information in Diagram Q10b, and also your knowledge of fair trade, **explain** how fair trade:

(i) helps to reduce inequalities in world trade; and

(ii) impacts on farming communities.

6

MARKS

Question 11: Energy

Diagram Q11: % Energy Production in Selected Countries

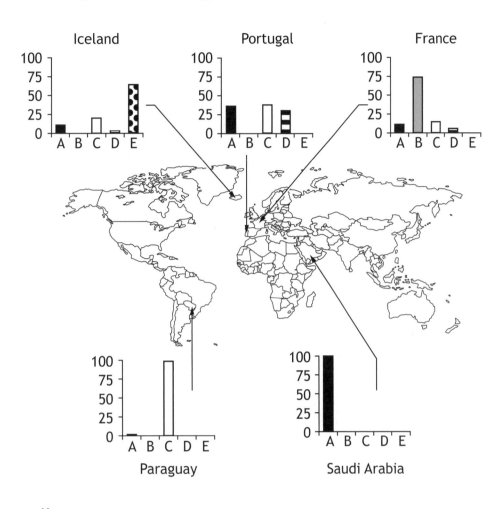

Key:

A ■ Fossil Fuels (oil, gas and coal)

B ▨ Nuclear

C ☐ Hydroelectric Power (HEP)

D ▤ Other Renewables (solar and wind power)

E ◨ Geothermal Power

Study Diagram Q11.

(a) Using the information in the diagram **suggest reasons** for the different patterns of energy production in the countries shown.   4

(b) Choose **one** renewable *and* **one** non-renewable approach to energy production, and for each approach, **evaluate** its effectiveness in meeting energy demands.   6

SECTION 4 : APPLICATION OF GEOGRAPHICAL SKILLS - 10 marks

Attempt the question

MARKS

Question 12

> There is a proposal to develop the site at 468057 on the OS Map extract as an Outdoor and Environmental Education centre at Dalmellington, East Ayrshire.
>
> The specifications required for this development are listed below.
>
> Specifications for an Outdoor & Environmental Education Centre:
>
> • offer a range of land-based activities — hillwalking, orienteering, mountain biking and abseiling/rock-climbing
>
> • provide facilities for fishing and water-sports
>
> • allow opportunities for environmental education, fieldwork and conservation in the local area, eg biology and geography studies.

Study Diagram Q12a – Location of Proposed Development; OS Map (Extract 2144/EXP327: Dalmellington), Diagram Q12b, Diagram Q12c, Diagram Q12d and Diagram Q12e before answering this question.

Referring to map evidence from the OS Map extract, and other information from the sources, **discuss**:

(a)   the advantages **and** disadvantages of the proposed location; **and**

(b)   any possible impacts on the local area and East Ayrshire.

10

Diagram Q12a : Location of Proposed Development

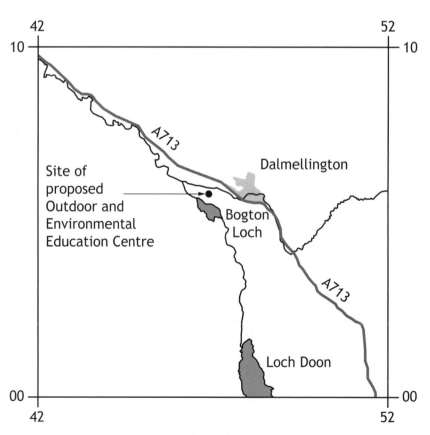

Question 12 (continued)

Diagram Q12b : Photograph of Bogton Loch & Dalmellington Moss looking SW from GR 466065

Dalmellington Moss is a Scottish Wildlife Trust Nature Reserve. Rare heathers and moss along with birds such as curlew and snipe can be found here.

Auchenroy
Hill

Bogton Loch is a Site of Special Scientific Interest because of the flora and fauna, including water birds, such as Teal and Reed Warblers, the Whooper swan and Greylag goose.

Diagram Q12c: Unemployment rates (%) for East Ayrshire and Scotland, 2008 - 2010

	2008	2009	2010
East Ayrshire	5·6	8·0	9·7
Scotland	4·5	5·9	7·6

Diagram Q12d: Population Change for Dalmellington and Burnton 2001 - 2010

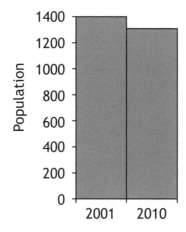

Diagram Q12e: Where tourists spend their money in Ayrshire and Arran

- South Ayrshire
- North Ayrshire and Arran
- East Ayrshire

[END OF QUESTION PAPER]

[BLANK PAGE]

DO NOT WRITE ON THIS PAGE

National
Qualifications
2015

X733/76/21

Geography
Ordnance Survey Map

THURSDAY, 21 MAY

9:00 AM – 11:15 AM

ORDNANCE SURVEY MAP

For Question 12

Note: The colours used in the printing of this map extract are indicated in the four little boxes at the top of the map extract. Each box should contain a colour; if any does not, the map is incomplete and should be returned to the Invigilator.

ROADS AND PATHS **Not necessarily rights of way**

M1 or A6(M)	Motorway 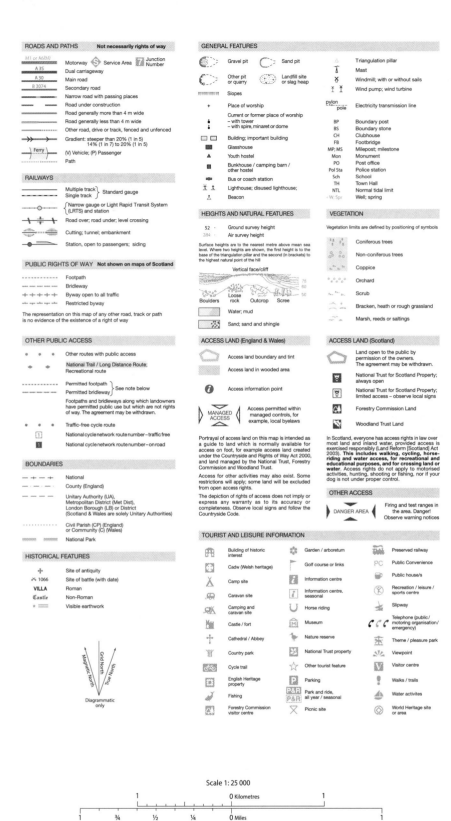 Service Area **7** Junction Number
A 35	Dual carriageway
A 30	Main road
B 3074	Secondary road
	Narrow road with passing places
	Road under construction
	Road generally more than 4 m wide
	Road generally less than 4 m wide
	Other road, drive or track, fenced and unfenced
	Gradient: steeper than 20% (1 in 5) 14% (1 in 7) to 20% (1 in 5)
Ferry	(V) Vehicle; (P) Passenger
	Path

RAILWAYS

	Multiple track } Standard gauge
	Single track
	Narrow gauge or Light Rapid Transit System (LRTS) and station
	Road over; road under; level crossing
	Cutting; tunnel; embankment
	Station, open to passengers; siding

PUBLIC RIGHTS OF WAY **Not shown on maps of Scotland**

– – – – – – –	Footpath
— — — — —	Bridleway
+ + + + + +	Byway open to all traffic
– + – + – + –	Restricted byway

The representation on this map of any other road, track or path is no evidence of the existence of a right of way

OTHER PUBLIC ACCESS

• • •	Other routes with public access
♦ ♦	National Trail / Long Distance Route; Recreational route
– – – – –	Permitted footpath } See note below
– – – – –	Permitted bridleway
	Footpaths and bridleways along which landowners have permitted public use but which are not rights of way. The agreement may be withdrawn.
• • •	Traffic-free cycle route
1	National cycle network route number – traffic free
1	National cycle network route number – on road

BOUNDARIES

– + – + –	National
– · – · – · –	County (England)
– – – –	Unitary Authority (UA), Metropolitan District (Met Dist), London Borough (LB) or District (Scotland & Wales are solely Unitary Authorities)
––––––	Civil Parish (CP) (England) or Community (C) (Wales)
▪▪▪▪ ▪▪▪▪	National Park

HISTORICAL FEATURES

╬	Site of antiquity
⚔ 1066	Site of battle (with date)
VILLA	Roman
Castle	Non-Roman
☩ ▦	Visible earthwork

Magnetic North — Grid North — True North

Diagrammatic only

GENERAL FEATURES

⬭	Gravel pit	⬭	Sand pit
⬭	Other pit or quarry	⬭	Landfill site or slag heap
�	Slopes		
+	Place of worship		
⛪	Current or former place of worship – with tower	△	Triangulation pillar
⛪	– with spire, minaret or dome	ⵣ	Mast
▢	Building; important building	ⵝ	Windmill; with or without sails
▦	Glasshouse	ⵝ ⵝ	Wind pump; wind turbine
▲	Youth hostel	pylon pole	Electricity transmission line
▣	Bunkhouse / camping barn / other hostel		
⬛	Bus or coach station	BP	Boundary post
ⵤ ⵤ	Lighthouse; disused lighthouse;	BS	Boundary stone
ⵣ	Beacon	CH	Clubhouse
		FB	Footbridge
		MP; MS	Milepost; milestone
		Mon	Monument
		PO	Post office
		Pol Sta	Police station
		Sch	School
		TH	Town Hall
		NTL	Normal tidal limit
		◁ W; Spr	Well; spring

HEIGHTS AND NATURAL FEATURES

52 ·	Ground survey height
284 ·	Air survey height

Surface heights are to the nearest metre above mean sea level. Where two heights are shown, the first height is to the base of the triangulation pillar and the second (in brackets) to the highest natural point of the hill

Vertical face/cliff

78
60
50

Boulders   Loose rock   Outcrop   Scree

	Water; mud
	Sand; sand and shingle

ACCESS LAND (England & Wales)

	Access land boundary and tint
	Access land in wooded area
ⓘ	Access information point
MANAGED ACCESS	Access permitted within managed controls, for example, local byelaws

Portrayal of access land on this map is intended as a guide to land which is normally available for access on foot, for example access land created under the Countryside and Rights of Way Act 2000, and land managed by the National Trust, Forestry Commission and Woodland Trust.

Access for other activities may also exist. Some restrictions will apply; some land will be excluded from open access rights.

The depiction of rights of access does not imply or express any warranty as to its accuracy or completeness. Observe local signs and follow the Countryside Code.

TOURIST AND LEISURE INFORMATION

⌂	Building of historic interest	✿	Garden / arboretum	🚂	Preserved railway
	Cadw (Welsh heritage)	⚑	Golf course or links	PC	Public Convenience
⛺	Camp site	ⓘ	Information centre	🍺	Public house/s
	Caravan site	ⓘ	Information centre, seasonal	⚽	Recreation / leisure / sports centre
	Camping and caravan site	⛎	Horse riding	⚓	Slipway
⛫	Castle / fort	Ⓜ	Museum	☎ ☎ ☎	Telephone (public / motoring organisation / emergency)
†	Cathedral / Abbey		Nature reserve		Theme / pleasure park
	Country park		National Trust property	⋇	Viewpoint
	Cycle trail	☆	Other tourist feature	Ⓥ	Visitor centre
	English Heritage property	P	Parking	!	Walks / trails
	Fishing	P&R P&R	Park and ride, all year / seasonal	⛵	Water activites
	Forestry Commission visitor centre	✕	Picnic site		World Heritage site or area

VEGETATION

Vegetation limits are defined by positioning of symbols

	Coniferous trees
	Non-coniferous trees
	Coppice
	Orchard
	Scrub
	Bracken, heath or rough grassland
	Marsh, reeds or saltings

ACCESS LAND (Scotland)

	Land open to the public by permission of the owners. The agreement may be withdrawn.
	National Trust for Scotland Property; always open
	National Trust for Scotland Property; limited access – observe local signs
	Forestry Commission Land
	Woodland Trust Land

In Scotland, everyone has access rights in law over most land and inland water, provided access is exercised responsibly (Land Reform [Scotland] Act 2003). **This includes walking, cycling, horse-riding and water access, for recreational and educational purposes, and for crossing land or water.** Access rights do not apply to motorised activities, hunting, shooting or fishing, nor if your dog is not under proper control.

OTHER ACCESS

DANGER AREA	Firing and test ranges in the area. Danger! Observe warning notices

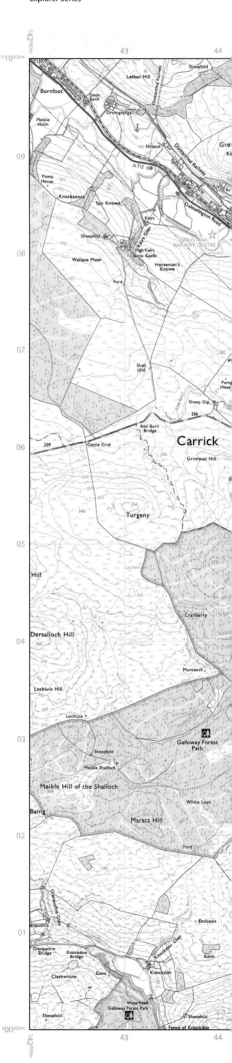

Scale 1: 25 000

1   0 Kilometres   1

1   ¾   ½   ¼   0 Miles   1

1 kilometre = 0·6214 mile   4 centimetres to 1 kilometre (one grid square)   1 Mile = 1·6093 kilometres

HIGHER FOR CfE | ANSWER SECTION

SQA AND HODDER GIBSON HIGHER FOR CfE GEOGRAPHY 2015

General Marking Principles for Higher Geography

This information is provided to help you understand the general principles you must apply when marking candidate responses to questions in this Paper. These principles must be read in conjunction with the detailed marking instructions, which identify the key features required in candidate responses.

a) Marks for each candidate response must always be assigned in line with these General Marking Principles and the Detailed Marking Instructions for this assessment.

b) Marking should always be positive. This means that, for each candidate response, marks are accumulated for the demonstration of relevant skills, knowledge and understanding: they are not deducted from a maximum on the basis of errors or omissions.

c) Where the candidate violates the rubric of the paper and answers two parts in one section, both responses should be marked and the better mark recorded.

d) Marking must be consistent. Never make a hasty judgement on a response based on length, quality of hand writing or a confused start.

e) Use the full range of marks available for each question.

f) The Detailed Marking Instructions are not an exhaustive list. Other relevant points should be credited.

g) For credit to be given, points must relate to the question asked. Where candidates give points of knowledge without specifying the context, these should be rewarded unless it is clear that they do not refer to the context of the question.

h) For knowledge/understanding marks to be awarded, points must be:
   a. relevant to the issue in the question
   b. developed (by providing additional detail, exemplification, reasons or evidence)
   c. used to respond to the demands of the question (ie evaluate, analyse, etc)

Marking principles for each question type

There are a range of types of question which could be asked within this question paper. For each, the following provides an overview of marking principles, and an example for each.

Explain

Questions which ask candidates to explain or suggest reasons for the cause or impact of something, or require them to refer to causal connections and relationships: candidates must do more than describe to gain credit here.

Where this occurs in a question asking about a landscape feature, candidates should refer to the processes leading to landscape formation.

Where candidates are provided with sources, they should make use of these and refer to them within their answer for full marks.

Where candidates provide a purely descriptive answer, or one where development is limited, no more than half marks should be awarded for the question.

Other questions look for higher-order skills to be demonstrated and will use command words such as analyse, evaluate, to what extent does, discuss.

Analyse

Analysis involves identifying parts, the relationship between them, and their relationships with the whole. It can also involve drawing out and relating implications.

An analysis mark should be awarded where a candidate uses their knowledge and understanding/a source, to identify relevant components (eg of an idea, theory, argument, etc) and clearly show at least one of the following:

- links between different components
- links between component(s) and the whole
- links between component(s) and related concepts
- similarities and contradictions
- consistency and inconsistency
- different views/interpretations
- possible consequences/implications
- the relative importance of components
- understanding of underlying order or structure

Where candidates are asked to analyse they should identify parts of a topic or issue and refer to the interrelationships between, or impacts of, various factors, eg analyse the soil-forming properties which lead to the formation of a gley soil. Candidates would be expected to refer to how the various soil formatting properties contributed to the formation.

Evaluate

Where candidates are asked to evaluate, they should be making a judgement of the success, failure, or impact of something based on criteria. Candidates would be expected to briefly describe the strategy/project being evaluated before offering an evidenced conclusion.

Account for

Where candidates are being asked to account for, they are required to give reasons, often (but not exclusively) from a resource, eg for a change in trade figures, a need for water management, or differences in development between contrasting developing countries.

Discuss

These questions are looking for candidates to explore ideas about a project, or the impact of a change. Candidates will be expected to consider different views on an issue/argument. This might not be a balanced argument, but there should be a range of impacts or ideas within the answer.

To what extent

This asks candidates to consider the impact of a management strategy or strategies they have explored. Candidates would be expected to briefly describe the strategy/project being evaluated before offering an evidenced conclusion. Candidates do not need to offer an overall opinion based on a variety of strategies, but should assess each separately.

Marking Instructions for each question

HIGHER FOR CfE GEOGRAPHY
SPECIMEN QUESTION PAPER

Section 1: Physical Environments

Question		General Marking Instructions for this type of question	Max mark	Specific Marking Instructions for this question
1.		Check any diagram(s) for relevant points not present in the text and award accordingly.	5	*Possible answers might include:*

General Marking Instructions for this type of question:

Check any diagram(s) for relevant points not present in the text and award accordingly.

Well-annotated diagrams that explain conditions and processes can gain full marks.

Maximum of 1 mark for undeveloped conditions and 1 mark for undeveloped processes.

Award a maximum of 2 marks for fully developed processes.

Answers must have conditions and processes to gain maximum marks.

Answers which are purely descriptive, or do not develop any processes or conditions, should achieve no more than 2 marks in total.

1 mark
Limited explanation – the use of the names of at least two processes with no development of these.

2 marks
The use of the names of at least two processes/conditions with development of these, but no other reference to conditions.

OR

Limited use of the names of at least two processes/conditions, with at least two descriptive points about the landscape formation.

3 marks
The use of the processes of at least two conditions with development of these, or two developed processes with limited explanation of how the feature forms over time.

Limited use of the names of at least three processes/conditions, with at least three descriptive points about the landscape formation.

4 marks
The use of the processes of at least three processes/conditions with development of these, or two developed processes, with two further statements explaining the formation of the feature.

Max mark: 5

Specific Marking Instructions for this question:

Possible answers might include:

- Snow accumulates in mountain hollows when more snow falls in winter than melts in the summer. (1 mark)
- North/north-east facing slopes are more shaded so snow lies longer (1 mark), with accumulated snow compressed into neve and eventually ice. (1 mark)
- Plucking, when ice freezes on to bedrock, pulling loose rocks away from the backwall, making it steeper. (1 mark)
- Abrasion, when the angular rock embedded in the ice grinds the hollow, making it deeper. (1 mark)
- Frost shattering continues to steepen the sides of the hollow when water in cracks in the rock turns to ice when temperatures drop below freezing; expansion and contraction weakens the rock until fragments break off. (1 mark)
- Rotational sliding further deepens the central part of the hollow floor as gravity causes the ice to move. (1 mark)
- Friction causes the ice to slow down at the front edge of the corrie, allowing a rock lip to form, which traps water as ice melts, leaving a lochan or tarn. (1 mark)
- During spring/summer, thawing takes place, allowing water to penetrate cracks in the rocks at the base of the hollow. (1 mark) The broken fragments build up over time and are removed by meltwater, further enlarging the hollow. (1 mark) Frost shattering on the backwall supplies further abrasion material as loose scree falls down the bergschrund. (1 mark)
- This is a large crevasse separating moving ice from the ice still attached to the backwall. (1 mark)

A weaker answer may take the form of descriptive points such as:

- A corrie forms when a glacier forms in a hollow and moves downhill, eroding an armchair shape. (1 mark) Plucking, abrasion and frost shattering help to erode the corrie. (1 mark) The ice steeped the back wall and deepens the hollow. (1 mark)

1.		(continued) **5 marks** The use of the processes of at least three processes/conditions with development of these, or three developed processes, with three further statements explaining the formation of the feature, including a named example.		
2.		Candidates need not refer to all factors in diagram for full marks – at least two factors are expected. "Explain" questions should make reference to *causal* relationships. Marks may be awarded as follows: • For 1 mark, candidates may give one detailed explanation, or a limited description/explanation of two factors. • Candidate responses may be a mixture of the two styles of writing, but a maximum of 3 marks should be awarded for answers entirely consisting of limited descriptive/explanatory points.	6	*Possible answers might include:* • Natural vegetation – deciduous forest vegetation provides deep leaf litter, which is broken down rapidly in mild/warm climate. (1 mark) • Trees have roots which penetrate deep into the soil, ensuring the recycling of minerals back to the vegetation. (1 mark) • Soil organisms – soil biota break down leaf litter producing mildly acidic mull humus. They also ensure the mixing of the soil, aerating it and preventing the formation of distinct layers within the soil. (1 mark) • Climate – precipitation slightly exceeds evaporation, giving downward leaching of the most soluble minerals and the possibility of an iron pan forming, impeding drainage. (1 mark) • Rock type – determines the rate of weathering, with hard rocks such as schist taking longer to weather, producing thinner soils. Softer rocks, eg shale, weather more quickly. (1 mark) • Relief – greater altitude results in temperatures and the growing season being reduced and an increase in precipitation. (1 mark) Steeper slopes tend to produce thinner soils due to gravity. (1 mark) • Drainage – well drained with throughflow and little accumulation of excess water collecting, producing limited leaching. (1 mark) Candidates may give a developed explanations with interactions between factors, for example: • The A horizon is rich in nutrients, caused by the relatively quick decomposition of the litter of deciduous leaves and grasses in a mild climate. (1 mark) • This produces a mull humus, well mixed with the soil minerals thanks to the activity of organisms such as worms. (1 mark) • Soil colour varies from black humus to dark brown in A horizon to lighter brown in B horizon where humus content is less obvious. Texture is loamy and well-aerated in the A horizon but lighter in the B horizon. (1 mark) • The C horizon is derived from a range of parent material, with limestone producing lighter-coloured alkaline soils. (1 mark) • South-facing slopes with a greater amount of sunshine and higher temperatures increase the rate of humus production. (1 mark)

| 3. | Check any diagram(s) for relevant points not present in the text and award accordingly.

 Well-annotated diagrams, explaining reasons, can gain full marks.

 • For 1 mark, candidates may give one detailed explanation, or a limited description/explanation of two factors.

 • Candidate responses may be a mixture of the two styles of writing, but a maximum of 2 marks should be awarded for answers entirely consisting of limited descriptive/explanatory points. | 4 | *Possible answers might include:*

 • Sun's angle in the sky decreases towards the poles due to the curvature of the Earth, which spreads heat energy over a larger surface area. (1 mark)
 • Sun's rays are concentrated on tropical latitudes as the intensity of insolation is greatest where rays strike vertically. (1 mark)
 • Sun's rays have less atmosphere to pass through at the tropics, so less energy is lost through absorption and reflection by clouds, gas and dust. (1 mark)
 • Albedo rates differ from the darker forest surfaces at the tropics absorbing radiation, in contrast to the ice-/snow-covered polar areas reflecting radiation. (1 mark)
 • Tilt of the axis results in the Sun being higher in the sky between the tropics throughout the year, focusing energy. (1 mark)
 • No solar insolation at the winter solstices at the poles producing 24-hour darkness, whereas the tropics receive insolation throughout the year. (1 mark) |

Section 2: Human Environments

Question	General Marking Instructions for this type of question	Max mark	Specific Marking Instructions for this question
1.	Candidates must explain the problems of collecting accurate population data in developing countries. No marks for description. Marks may be awarded as follows: • For 1 mark, candidates may give one detailed explanation, or a limited description/explanation of two factors. Detail may include relevant exemplification of a problem. • Candidate responses may be a mixture of the two styles of writing, but a maximum of 3 marks should be awarded for answers entirely consisting of limited descriptive/explanatory points.	6	*Possible answers might include:* • Large numbers of migrants, eg the Tuareg or Fulani in West Africa, and the shifting cultivators of the Amazon, may lead to people being missed or counted twice. (1 mark) • Countries with large numbers of homeless people or large numbers of rural-to-urban migrants living in shanty towns, eg Makoko in Lagos, Nigeria, have no official address for an enumerator to visit. (1 mark) • Poor communication links and difficult terrain, eg in the Amazon Rainforest, may make it difficult for enumerators to reach isolated villages. (1 mark) • The variety of languages spoken in many countries (eg over 500 in Nigeria) make it difficult to provide forms that everyone can complete. (1 mark) • The considerable costs involved in printing, training enumerators, distributing forms and analysing the results can make conducting a census impossible, especially when the country may have more pressing problems like housing and education. (1 mark) • In countries with high levels of illiteracy, mistakes may be made and more enumerators will be needed to help. (1 mark) • People may be suspicious of why the census is being conducted, and may lie. (1 mark) • Ethnic tensions and internal political rivalries may lead to inaccuracies, eg northern Nigeria was reported to have inflated its population figures to secure increased political representation. (1 mark) • Under-registration may occur for social, religious and political reasons, eg China's one-child policy may have reduced the registration of baby girls. (1 mark) • In countries suffering from war, eg Afghanistan, it may be dangerous for enumerators to enter regions and data will quickly become dated. (1 mark)

| 2. | | Candidates should analyse the impact of the migration on either the donor or the recipient country. Advantages and disadvantages must be included for full credit.

Markers should take care not to credit reversed points.

Answers will depend on the case study described by the candidate.

Marks may be awarded as follows:

• For 1 mark, candidates may give one detailed explanation, or a limited description/explanation of two factors. Detail may include relevant exemplification of a problem.

• Candidate responses may be a mixture of the two styles of writing, but a maximum of 2 marks should be awarded for answers entirely consisting of limited descriptive/explanatory points.

A maximum of 2 marks should be awarded for answers which are vague or over-generalised. | 5 | *Possible answers might include:*

Donor country, eg Greece, Spain or Bulgaria:
Advantages
• Pressure on local services such as education, healthcare and housing is reduced. (1 mark)
• Pressure on jobs is reduced therefore levels of unemployment will fall. (1 mark)
• The birth rate is lowered so population growth rates will slow. (1 mark)
• Money sent home by the migrants will boost the local economy. (1 mark)
• Migrants will learn new skills and may then return to their home country. (1 mark)

Disadvantages
• Active and most educated population left, known as the 'brain drain', which resulted in a skills shortage in donor countries. (1 mark)
• Families were divided and the death rates may increase due to the ageing population. (1 mark)
• Family members remaining in the country of origin may become dependent on remittances being sent home by migrant workers. (1 mark)

Recipient country, eg Germany:
Advantages
• The short-term gap in labour is filled. Many migrants are highly skilled, eg engineers and academics. (1 mark)
• Migrants will take jobs that locals did not want and will work for lower, more competitive wages, thus reducing labour costs. (1 mark)
• Migrants will enrich the culture of the area that they move to with language, food and music. (1 mark)
• The increased population will result in an increase in the tax paid to the government, which can be invested in improving local services. (1 mark)

Disadvantages
• Migrant workers may feel discriminated against. Unemployment rises for local people. (1 mark)
• Ghettos may develop in parts of cities and there may be a shortage of affordable housing. (1 mark)
• Cost of providing services for migrant population and their families will increase, eg for schooling, healthcare, etc. (1 mark) |
| 3. | | Answers will depend on the case study referenced by the candidate.

Marks may be awarded as follows:

• For 1 mark, candidates may give one detailed explanation, or a limited description/explanation of two factors. Detail may include relevant exemplification of a problem.

• Candidate responses may be a mixture of the two styles of writing, but a maximum of 2 marks should be awarded for answers entirely consisting of limited descriptive/explanatory points.

A maximum of 2 marks should be awarded for answers which are vague or over-generalised. | 4 | *Possible answers might include:*

Eden Foundation in Nigeria:
• Educated farmers to grow perennial plants to protect the soil against heavy rain. (1 mark)
• They prevent rainsplash from dislodging fine particles and bind the loose soil. (1 mark)
• Farmers produced twice as much millet (drought-tolerant crop) as those who did not use this technique. (1 mark)
• Undisturbed by ploughing, the soil structure will remain intact. (1 mark) Organic matter holds the soil particles together.
• Stone lines are commonly used in Burkina Faso and Niger, to trap soil and water, and slow run-off. (1 mark)
• Instead, water will sink into the soil through the cracks and pours, preventing erosion. (1 mark)
• Strips of vegetation can also be used in a similar way, and can provide fodder for animals (eg Makarikari grass) or a cash crop of pumpkins could be grown. (1 mark)
• Build wells to allow effective irrigation. (1 mark)
• Contour ridges slow run-off and catch sediment before it is washed away. (1 mark) |

| 3. | | (continued) | | **Afforestation**
• To prevent soil erosion as roots will bind the soil and hold it in place. (1 mark)
• Fanya juu terraces (popular in Makanya in north-eastern Tanzania) can be made by digging a drainage channel and throwing soil uphill to make a ridge. In drier areas, trees can be planted in the ditch, and in wetter areas on the ridge. (1 mark)
• In Makanya, maize is grown between the trenches. Maize crops have increased from 1.5 tonnes per hectare to 2.4 tonnes per hectare. (1 mark) |

Section 3: Global Issues

River Basin Management

Question		General Marking Instructions for this type of question	Max mark	Specific Marking Instructions for this question
1.	(a)	Candidates must explain the need for water management in Ghana using the relevant prompts from the resources. No marks should be awarded for purely descriptive points, eg 'very low rainfall in Ghana during the winter months'. Marks may be awarded as follows: • For 1 mark, candidates may give one detailed explanation, or a limited description/explanation of two factors. • Candidate responses may be a mixture of the two styles of writing, but a maximum of 2 marks should be awarded for answers which are purely descriptive.	5	*Acceptable reasons might include:* • Very low rainfall in the north of Ghana from November to March means that water management is needed to ensure that people have water for washing/drinking, etc, all year round. (1 mark) • High rainfall in the summer months throughout Ghana shows that flooding is a threat. Building dams to allow flood control is required. (1 mark) • The rapid population growth predicted in the future for Ghana suggests that demand for water from the growing population will be high and increasing (for all the usual uses of water), and this means water management is required. (1 mark) • Most people in Ghana (56%) are employed in agriculture and this shows that water management is important for irrigation purposes. (1 mark) • Only 45% of people in Ghana have access to electricity, and as their respective population grows the importance of hydroelectric power is evident. (1 mark) • Ghana's capital city of Accra is not situated on the main river (the Volta) and so it needs to be managed to allow it to benefit from domestic and industrial water use. (1 mark)
	(b)	Answers must discuss the possible negative consequences. No marks should be awarded for positive consequences. Although no mark is to be awarded for the named water management project, vague/generic answers that do not relate to a specific named water management project should get a maximum of 4 marks. Candidates who only deal with socio-economic or environmental impacts should be awarded a maximum of 4 marks. For 1 mark, candidates may give one detailed explanation, or a limited description/explanation of two factors.	5	*Answers will depend on the water management project chosen, but for Ghana, possible answers might include:* **Environmental consequences could include:** • Flooding of animal habitat, eg 21% of the Bui National Park has been flooded, with fears that the rare black hippopotamus may have been threatened due to insufficient suitable habitats near the inundated area. (1 mark) • The moderation of river flow downstream of the dams could adversely affect fish habitats — 46 species of fish could be adversely affected by the Bui Dam changing the water flow, temperature and turbidity and blocking migration routes. (1 mark) • Rotting vegetation in the new lakes may release greenhouse gases, which could increase climate change. (1 mark)

1.	(b)	**(continued)** Candidate responses may be a mixture of the two styles of writing, but a maximum of 2 marks should be awarded for answers which are purely descriptive. 1 mark can be awarded where candidates refer to **two** specific named examples *within* the case study area.		**Socio-economic consequences could include:** • The forcible displacing of people. In building the Akosombo Dam an estimated 80,000 people were displaced and relocated into resettlement villages (1 mark) but many of these villages were not capable of providing to the same level of income as villages previously had (with poorer soils). (1 mark) • The increased spread of diseases (eg Schistosomiasis and malaria), which have been linked to the creation of the stagnant lake. (1 mark) In addition to this, there has been an increase in the incidence of AIDS in the Volta Basin communities, linked to the increased ease of migration since the creation of Lake Volta. (1 mark) • A decline in agricultural productivity in the area surrounding Lake Volta as the soils here are less fertile than the soils now submerged under the lake. (1 mark) In addition, without the natural river floods to replace nutrients, there has been increased chemical use and the lake is now suffering from eutrophication. (1 mark) This invasion of river weeds is making fishing and navigation by motor boat across Lake Volta more difficult. (1 mark)

Development and Health

Question	General Marking Instructions for this type of question	Max mark	Specific Marking Instructions for this question
2.	Candidates may choose to answer parts (a) and (b) separately or together. Award a maximum of 6 marks for either section. Candidates must explain how each method actually helps to control the disease and not just describe or list different methods, eg releasing natural predators into the environment, such as Nile tilapia which eat the larvae. Evaluation points should also be developed points for a mark to be awarded, eg is relatively cheap, therefore affordable for developing world countries. Each evaluation should only be credited once — ie candidates should be credited for, eg, cost only once. Care should be taken not to credit reversals. For 1 mark, candidates may give one detailed explanation, or a limited description/explanation of two factors. Candidate responses may be a mixture of the two styles of writing, but a maximum of 2 marks should be awarded for answers which are purely descriptive. 1 mark can be awarded where candidates refer to **two** specific named examples *within* the case study area, up to a maximum of 2 marks.	10	*Possible answers for methods might include:* • The female anopheles mosquito acts as a vector for the transmission of malaria, so one method used was to spray pesticides/insecticides such as DDT in an attempt to kill the mosquitoes by destroying their nervous systems. (1 mark) • Breeding genetically modified sterile mosquitoes and mercenary male mosquitoes were also attempts to kill off the mosquito for good, and so stop the spreading of the disease. (1 mark) • Another method was to use specially designed mosquito traps, which mimic animals and humans by emitting a small amount of carbon dioxide in order to lure the mosquitoes into the trap where they are killed. (1 mark) • BTI bacteria can be artificially grown in coconuts and then, when the coconuts are split open and placed in a stagnant pond, the larvae eat the bacteria which destroy the larvae stomach lining, killing them. (1 mark) • Putting larvae-eating fish such as the muddy loach into stagnant ponds or paddi fields can also help to reduce the larvae as the fish eat the larvae. (1 mark) • Other methods were aimed at getting rid of the stagnant water required for mosquitoes to lay their eggs, eg draining stagnant ponds or swamps every seven days as it takes longer than this period of time for the larvae to develop into adult mosquitoes. (1 mark) • Planting eucalyptus trees, which soak up excess moisture in marshy areas, was also an attempt to prevent the formation of stagnant pools. (1 mark) • Covering water storage cans/small ponds was also used as an attempt to stop mosquitoes from reproducing successfully. (1 mark) • The increased use of insecticide-coated mosquito nets at night was an attempt to stop the mosquitoes from biting people and passing on the disease as they slept. (1 mark)

2.		(continued)		• Attempts were also made to cure people once they had contracted the disease by killing the plasmodium parasite once people had been contaminated with it. Drugs like Quinine, Chloroquine, Larium and Malarone were all developed in an attempt to kill the parasite. (1 mark) • A drug developed from the Chinese herb Artemisia, and an artificial version of this called 'Oz', appears to work in some parts of the world at least, by reacting violently with the iron in the parasite and killing it before the parasite can adapt. (1 mark)
				Possible comments on the effectiveness might include: • Insecticides to kill the mosquito were effective at first and helped to eradicate the disease in Southern Europe and Florida, however the mosquito became resistant to DDT and alternative insecticides are often too expensive for developing countries. (1 mark) • Mosquito traps have been effective at a small scale, but mosquitoes breed so quickly that it is impossible to trap them all. (1 mark) • The approaches aimed at killing the mosquito larvae have had only limited success (and only at a local scale) and have been criticised for causing pollution/changing the ecosystem of water courses. (1 mark) • The BTI bacteria in coconuts is a cheap and environmentally friendly solution, with 2/3 coconuts clearing a typical pond of mosquito larvae for 45 days. (1 mark) • Draining stagnant ponds is impossible to be effective on a large-scale, especially in tropical climates where it can rain heavily most days. (1 mark) • Using mosquito nets at night/covering up exposed skin is effective as mosquitoes are often most active during dusk and dawn. (1 mark) • Drugs to kill the parasite once inside humans have been effective for a spell, but the parasite often adapts and becomes resistant — this is true even of the Artemisia-based drugs in SE Asia. (1 mark) • Anti-malarial drugs often have unpleasant side-effects such as nausea, headaches and in some cases hallucinations. (1 mark) • They are also expensive to research, develop and produce, making them often too expensive for people living in developing countries. (1 mark) • Attempts are ongoing to develop a vaccine that could eradicate malaria for good, but so far this has not been successful. (1 mark)

Global Climate Change

Question		General Marking Instructions for this type of question	Max mark	Specific Marking Instructions for this question
3.	(a)	Award a maximum of 1 mark for an explanation of the greenhouse effect. 1 mark should be awarded for a source along with an explanation of why it is increasing, eg methane is released from cattle's digestive system and beef is increasingly in demand across the world. A maximum of 2 marks should be awarded for a list of sources of individual gases.	5	*Possible answers might include:* • Carbon dioxide from burning fossil fuels — road transport, power stations, heating systems, cement production — and from deforestation, particularly in the rainforests where more carbon dioxide is present in the atmosphere and less being recycled in photosynthesis (1 mark) and peat bog reclamation/development (particularly in Ireland and Scotland for wind farms). (1 mark) • CFCs: disused refrigerators release CFCs when the foam insulation inside them is shredded. (1 mark) The coolants used in fridges and air conditioning systems create CFCs which are safe in a closed system, but can be released if appliances are not disposed of correctly. (1 mark)

3.	(a)	(continued)		• Methane: from rice paddies to feed rapidly increasing populations in Asian countries such as India and China (1 mark), belching cows to meet increasing global demand for beef. (1 mark) Methane released from permafrost melting in Arctic areas due to global warming. (1 mark) • Nitrous oxides: from vehicle exhausts and power stations. (1 mark) • Sulphate aerosol particles and aircraft contrails: global 'dimming' — increase in cloud formation increases reflection/absorption in the atmosphere and therefore cooling. (1 mark)
	(b)	For 1 mark, candidates may give one detailed explanation, or a limited description/explanation of two factors. Candidate responses may be a mixture of the two styles of writing, but a maximum of 2 marks should be awarded for answers which are purely descriptive. 1 mark can be awarded where candidates refer to **two** specific named examples (species, ocean currents, or use of numeracy) *within* the case study area.	5	*Possible answers might include:* • Rise in sea levels caused by an expansion of the sea as it becomes warmer and also by the melting of glaciers and ice caps in Greenland, Antarctica, etc. (1 mark) • Low-lying coastal areas, eg Bangladesh affected with large-scale displacement of people and loss of land for farming and destruction of property. (1 mark) • More extreme and more variable weather, including floods, droughts, hurricanes, tornadoes becoming more frequent and intense. (1 mark) • Globally, an increase in precipitation, particularly in the winter in northern countries such as Scotland, but some areas like the USA Great Plains may experience drier conditions. (1 mark) • Increase in extent of tropical diseases, eg yellow fever as warmer areas expand, possibly up to 40 million more in Africa being exposed to risk of contracting malaria. (1 mark) • Longer growing seasons in many areas in northern Europe for example, increasing food production and range of crops being grown. (1 mark) • Impact on wildlife, eg extinction of at least 10% of land species and coral reefs suffer 80% bleaching. (1 mark) • Changes to ocean current circulation, eg in the Atlantic the thermohaline circulation starts to lose impact on north-western Europe, resulting in considerably colder winters. (1 mark) • Changes in atmospheric patterns linking to monsoon, El Nino, La Nina, etc. (1 mark)

Trade, Aid and Geopolitics

Question		General Marking Instructions for this type of question	Max mark	Specific Marking Instructions for this question
4.	(a)	Candidates should explain the inequalities in world trade patterns. For 1 mark, candidates may give one detailed explanation, or a limited description/explanation of two factors. Candidate responses may be a mixture of the two styles of writing, but a maximum of 3 marks should be awarded for answers which are purely descriptive.	5	*Possible answers might include:* • Developing countries often sell primary products at low value, therefore profits are limited. (1 mark) • Often, many countries are producing the same raw material, which keeps prices low. (1 mark) • However, developed countries manufacture products, which adds value and provides increased profits. (1 mark) • Developing countries are prevented from setting up coffee processing plants as high import taxes would be placed on processed coffee, whilst developed countries import coffee beans. (1 mark) • Patterns established during colonial times have been difficult to break. (1 mark) • Limits and quotas are also enforced, eg Kenya's export of coffee to the European Union is subject to a tariff of 9%, whilst other countries are subject to a 3.1% tariff. (1 mark)

4.	(a)	(continued)		• Developing countries are often very dependent on one or two products, eg bananas, sugar or copper. (1 mark) • Developed countries set the prices for raw materials through trading on commodity exchanges around the world, eg the New York Mercantile Exchange. (1 mark)
	(b)	Candidates should explain the effectiveness of strategies to reduce inequalities in world trade. For 1 mark, candidates may give one detailed explanation, or a limited description/explanation of two factors. Candidate responses may be a mixture of the two styles of writing, but a maximum of 2 marks should be awarded for answers which are purely descriptive. 1 mark can be awarded where candidates refer to **two** specific named examples *within* the case study area.		*Possible answers might include:* • The World Trade Organisation, established in 1996, settles trade disputes and continues to promote free trade and the removal of tariffs and quotas. (1 mark) • The removal of trade barriers means that developing countries will have access to lucrative markets in the developed world. (1 mark) • Some countries, eg in the Caribbean, are attempting to diversify their trade by developing non-traditional exports such as new crops or manufactured goods. Others are pursuing new markets. (1 mark) • The creation of trade alliances: the Caribbean Community and Common Market (CARICOM) was established to promote trade between Caribbean countries. (1 mark) • Customs duties between member states were removed, thus even the smallest Caribbean countries have access to a regional market. (1 mark) • Member countries are encouraged to purchase raw materials from other CARICOM countries. This has spread the benefits of industrialisation and has encouraged industries to locate in the smaller countries. (1 mark) • Within CARICOM, the Organisation of Eastern Caribbean States (OECS) has been established, which groups together the seven smallest countries in terms of their population. (1 mark) • OECS has created a single currency which makes trade between OECS countries much easier as money is not lost in transactions. (1 mark) • The OECS developed a common agricultural policy to subsidise farmers and removed controls on the movement of workers, allowing skilled workers to migrate. (1 mark) • Fairtrade guarantees a fair price for produce which always covers the cost of production regardless of the market price. (1 mark). • Five-year rolling contracts can be given which allows long-term planning to take place for investment in farm machinery, education, etc. (1 mark)

Energy

Question		General Marking Instructions for this type of question	Max mark	Specific Marking Instructions for this question
5.	(a)	No marks for describing the differences — candidates must account for the differences.	4	Although at present the amount of energy used by 'developed countries' is slightly higher than the amount used by 'developing countries', this is forecast to change. In the next 20 years the energy use in 'developing countries' is expected to increase at a much faster rate, whereas the 'developed countries' rate is expected to be stable, only rising slowly. Why is this? *Possible answers might include:* • Most of the global economic and population growth is happening in the 'developing' countries. (1 mark) This is causing an increase in demand from: o **Residential use** — with increased prosperity comes an increase in the standards of living for billions of people. Electricity for lighting and appliances such as televisions, washing machines, air conditioning, etc, will all cause energy use to increase. (1 mark)

5.	(a)	(continued)		o **Industrial use** — unlike in 'developed' countries, much of the economic growth in 'developing' countries is based on energy-hungry manufacturing industries. This accounts for some of the increased energy use here. (1 mark) o **Transport** — in a global economy many of the manufactured products are sold to 'developed' countries, and therefore need to be transported around the world — using energy. (1 mark) As people in 'developing' countries become more prosperous, car ownership rates will also increase, causing more energy use. (1 mark) • In 'developed' countries the population growth rates are more stable (or even declining), and so there is not any great increase in demand for energy. (1 mark) New products and technologies are also increasingly more energy-efficient which keeps energy consumption here more steady. (1 mark)
	(b)	Candidates must comment on the suitability of each renewable approach discussed. Care should be taken not to credit reverse statements twice (eg solar energy is effective in Spain because it has long hours of sunshine, whereas it is less effective in Scotland where there are fewer hours of sunshine). For 1 mark, candidates may give one detailed explanation, or a limited description/explanation of two factors. Candidate responses may be a mixture of the two styles of writing, but a maximum of 3 marks should be awarded for answers which are purely descriptive. 1 mark can be awarded where candidates refer to **two** specific named relevant examples.	6	*Possible answers might include:* • **Hydroelectric power** is more effective where there is high rainfall to ensure that reservoirs are always at capacity (1 mark), and suitable underlying geology (impermeable rock) to ensure water is not lost from the reservoirs through seepage. (1 mark) Hanging valleys often make ideal sites for effective hydroelectric power because they allow the vertical drop of water needed to power turbines. (1 mark) Hydroelectric pump storage schemes allow electricity production to be instantly produced as required, and are currently being used to meet periods of peak demand in Scotland. (1 mark) • **Wind power** is most effective where there are no barriers to the prevailing wind to allow regular and reliable movement of air to turn the turbines. (1 mark) Concerns have been raised about how to bridge the energy gap when the wind turbines are not generating electricity on calm days, because it is difficult to store the electricity produced from wind turbines. (1 mark) Other drawbacks to this approach, for developing countries, focus on the initial high costs of construction, but after this the energy produced is cheap. (1 mark) • **Wave power** approaches are currently being developed, and are most effective in areas such as the Pentland Firth where the fetch is large giving powerful waves, and in areas such as Cornwall. • **Tidal power** is used in areas where there is a large tidal range to create tidal currents which can be used to turn turbines, eg in the Pentland Firth or Bay of Fundi. • **Solar energy** is most effective where there are long hours of intense sunshine (eg in Spain) to power the solar panels. (1 mark) • **Geothermal energy** is most effective in tectonically active zones (such as Iceland), where there is a heat source (from magma) closer to the surface of the Earth, which can be used to generate steam. (1 mark) It is a reliable source of energy and can be used as needed. (1 mark) • **Biomass energy and biofuels** can provide continuous energy as required by the burning of plant matter. (1 mark) Drawbacks include concerns about air pollution to the local area increasing, and using land that is needed for crops production in developing countries. (1 mark) As carbon dioxide released equals what the plants recently took in, biomass energy does not add new greenhouse gasses, and so is more environmentally friendly than burning fossil fuels. (1 mark)

Section 4: Application of Geographical Skills

Question	General Marking Instructions for this type of question	Max mark	Specific Marking Instructions for this question
1.	Candidates should make reference to all sources, including the OS map to evaluate the suitability of the route in relation to the brief. For 1 mark, candidates should refer to the resource and offer an explanation with reference to the brief, or a limited description/explanation of two factors. A maximum of 5 marks should be awarded for answers consisting solely of limited descriptive points. A maximum of 4 marks should be awarded for candidates who give vague over-generalised answers which make no reference to the map. There are a variety of ways for candidates to give map evidence including descriptions, grid references and place names. A maximum of 4 marks should be awarded for answers which make no evaluation of the plan.	10	*Possible answers might include:* **Suitable for all levels of runners:** • The roads and streets are narrower here, causing a 'bunch' start for the runners which may cause problems. (1 mark) • While the steep uphill section on the A15 and even steeper downhill section on the B1188 may cause some difficultly for runners (1 mark), the generally flat nature of the route is likely to allow for a quick time, encouraging runners. (1 mark) • The exposed area between the 4km and 7km marks may cause some difficulties with the weather conditions/wind. (1 mark) **Cause minimum disruption to people and business in the local area:** • In grid square 9769 and 9770 the route goes through a residential area — local people may be unhappy with the lack of access by road. (1 mark) • The litter/noise levels (bands on the run) from runners/spectators may also cause disruption in residential areas. (1 mark) • The route uses the A57 (dual carriageway), a major road — closure will cause increased congestion and disruption. (1 mark) • Having the run on a Sunday will minimise this disruption as the area/businesses are likely to be quieter. (1 mark) **Promote business in local area:** • The route starts and finishes in the CBD/historic centre of Lincoln, promoting the local area to runners/supporters. (1 mark) • The estimated number of runners with extra supporters could be beneficial for some shops near to the start/finish line. (1 mark) • Hotels and camp/caravan sites will have extra business due to the increase in visitors. (1 mark) • Holding the run in February would bring in visitors at a time when trade is traditionally low. (1 mark) **Provide access for runners to start line:** • Car parking and transport to the start line may be problematic due to the lack of space in the CBD. (1 mark) • The start/finish of the route is near to the train and bus station allowing for some of the runners to get there by public transport and avoid parking. (1 mark) **Scenic/interesting for runners:** • The start/finish is in the historic centre of Lincoln (Castle/Cathedral), which may be of interest to the runners. (1 mark) • There is a varied landscape going through rural and urban environments, which may be more scenic for the runners. (1 mark) **Improvements may include:** • Starting the race in Hartsholme country park/West Common race course is likely to allow for easier access by car/parking. (1 mark) Hartsholme country park would be more scenic, as there is more green space and lakes. Having the run in the summer months may attract a bigger number of runners/spectators and boost trade. (1 mark)

General Marking Information for the Model Papers

When answering a question, you should ensure that you read the question carefully before you start. You should look at the command word to ensure you answer the question correctly, eg if the question asks you to explain, then you must give reasons to support your response. Remember that there are no describe questions so, if your answers do not contain more than a descriptive point, it will be difficult to achieve any marks. If a question asks for two points of view to be covered, for example, advantages and disadvantages, then both must be covered for full marks. If the question is out of six and only the advantages are covered, then a maximum mark of four or five might be awarded. If a named area or example is asked for, then you may lose a mark(s) if you give a general response to the question. There are often more points covered (worth a mark each) in the following answers than will be required. For example, there could be six or seven points listed for a five-mark answer. Since the question is only worth five marks then any five of the six points could gain the five marks.

HIGHER FOR CfE GEOGRAPHY
MODEL PAPER 1

Section 1: Physical Environments

Question		Specific Marking Instructions for this question	Max mark
1.		Between the Equator and the tropics, the sun's rays have less atmosphere to travel through, so less energy is lost through absorption and reflection (1). The rays of the sun cover a smaller area so are more concentrated, therefore, the intensity of insolation is higher here (1). At the poles, the earth is tilting away from the sun so the sun's rays have to travel further through the atmosphere so energy covers a larger area, therefore, insolation is lower here (1). At the tropics, areas of dense vegetation like the rainforest absorb radiation, whereas at the poles areas are covered in snow, and ice reflects the incoming radiation back into the atmosphere (1). Rays have to spread out and travel through more atmosphere so are more diluted as the earth tilts away from the Sun (1).	5
2.		Fully annotated diagrams could achieve full marks. A headland is a piece of land which juts out into the sea and, where there are weaknesses in the rock, erosion takes place by processes including chemical weathering because of the salt in the water which helps corrode the cliff (1). The sheer force of the waves crashing against the cliffs can erode it; this is called hydraulic action, where air may become trapped in joints and cracks on a cliff face. When a wave breaks, the trapped air is compressed which weakens the cliff and causes erosion (1).	5
		The base of the cliff can become undercut through attrition, where stones and rocks are hurled against the rock wearing it away (1). Through time, the weaknesses in the rock become wider as the waves force their way into cracks in the cliff face. The water contains sand and other materials that grind away at the rock until the cracks become a cave (1). Continued hydraulic action by pounding waves on the roof of the cave forms a tube. Eventually, this breaks through the surface of the ground near the edge of the cliff. At high tide, incoming waves force water out of the top of the blowhole (1). Waves can attack the headland on both sides and erode back-to-back caves, which eventually meet, and an arch will form (1). During stormy weather or high tides, the archway will become weakened and eventually the roof cannot be supported and it collapses, leaving a tall pillar or rock isolated from the headland which is called a stack (1).	
3.		The removal of natural vegetation and replacement with impermeable concrete/hard surfaces and drains (1) can speed up overland flow and can lead to higher river levels (1). In urban areas, people remove trees and vegetation then cover soil in impermeable materials like tarmac or concrete which will increase surface run-off (1). This leads to higher river levels and increases the risk of flooding (1). It also reduces the amount of water which returns to groundwater storage and possibly reduces the water table (1). Deforestation means there are no tree leaves and roots to soak up precipitation, leading to increased run-off and the potential for soil erosion (1). Deforestation can lead to a decrease in evapo-transpiration rates which means less moisture going into the atmosphere so less cloud formation, and so less rainfall impacting on local rainfall patterns (1). Water removed from rivers and underground stores for irrigation results in reduced river flow and lowers the water table (1). The silting-up of lakes, rivers and reservoirs due to waste products and mining processes can result in reduced storage in these areas (1).	5

Section 2: Human Environments

Question		Specific Marking Instructions for this question	Max mark
1.		In countries like Nigeria, some politicians and government officials from certain regions alter the figures by adding people so that their region gets extra funds from the government budgets. This results in the census figures being higher than the actual population (1). The variety of languages spoken in many countries (eg over 500 in Nigeria) makes it difficult to provide forms which everyone can understand, so the form is either not filled in or completed inaccurately (1). A census is expensive, involving costs of printing, distributing and analysing results, which poor countries cannot afford, especially with increasing populations needing housing, health care and education (1). Rural–urban migration results in creation of shanty towns, eg Makoko in Lagos, Nigeria where many people have no address for the enumerator to deliver the census form to, or many people are homeless and sleep on streets moving frequently so large numbers of people are missed out (1). In the Amazon Basin, the local tribes are shifting cultivators. This may lead to people being missed or counted twice, and many cannot read and write (1). The dense rainforest has poor access, with few communication links. It's difficult for an enumerator to reach, so people do not receive forms (1). In countries like China with a one-child policy, people lie about the number of children they have because they are afraid they will lose benefits or have to pay extra taxes (1).	5
2.		**For full marks both people and the environment should be mentioned.** Removal of the forest for mining, logging etc destroys the way of life of the indigenous people as their hunting area may be destroyed, leaving them short of food and forcing them to move further inland (1). The tribes use the land in a sustainable manner but loggers and miners destroy the area, completely preventing it from regenerating (1). Clashes between various competing groups occur, eg the violent death of Choco Mendez allegedly at the request of ranchers (1). Reduction of the fallow period, leading to reduced yields, leaves the population with less food (1). The using up of tribal land resulted in the creation of reserves for the indigenous people, causing unrest and conflict with the developers (1). The new settlers brought disease which the local people had no immunity to, causing large numbers to die (1). Local farmers have been displaced and forced to move to crowded cities, where they end up living in favelas (1).	6
3.		**Marks may be lost for a generalised answer which does not refer to a specific city.** **If Glasgow chosen:-** Pedestrianisation of areas in the city centre to create traffic-free areas, eg Buchanan Street (1). Park-and-ride schemes are used, where people park their car on the outskirts and travel into the city by train or bus, to encourage the use of public transport, eg Merriton, Chatelherault (1). One-way systems, eg George Square, improve the flow of traffic because vehicles do not need to slow down to pass each other nor look for other vehicles coming in the opposite direction (1). Parking restrictions and making parking more expensive discourages motorists from using their car in the city centre, making the streets wider allowing traffic to flow more easily (1). Multi-storey car parks hold large amounts of cars, reducing congestion on the streets (1). Dedicated bus lanes reduce travelling time, making public transport more efficient and attractive and reducing the volume of cars (1). Improvements in the road system, with new links created to bypass congested areas, removing unnecessary traffic from narrow roads (1).	4

Section 3: Global Issues

Question		Specific Marking Instructions for this question	Max mark
1. River Basin Management			
	(a)	The Missouri's drainage basin has highly variable weather and rainfall patterns. Most of the drainage basin receives less than 250mm of rainfall per year, so it suffers from drought at certain times of the year, but some of the most Westerly areas of the basin in the Rockies may receive up to 1000mm causing flooding (1). Most of the rainfall happens in winter but the intense heat of summer causes violent thunderstorms, again leading to overland flow and flooding (1). The river has many tributaries, causing river levels to rise dangerously especially with the spring snowmelt (1). The river carries a massive amount of sediment, which prevents a clear view of the bottom, in the past causing many ships to be wrecked (1). There is not a dependable flow to maintain a navigation channel year-round, which could hamper trade and tourism (1). The increasing population means an increasing demand for a water supply, so, managing the river ensures a constant supply of water for drinking and domestic use (1).	5

Question		Specific Marking Instructions for this question	Max mark
	(b)	**For full marks both the people and the environment must be mentioned.** **If the Missouri chosen:** Control of the river, through the construction of levees along the lower river and major tributaries, channelisation of floodplain tributaries, and an extensive reservoir system in the large tributary basins of the Platte, Kansas, and Osage rivers, stopped flooding in many areas, resulting in less people being displaced and less injuries/deaths (1). The six main dams and their reservoirs can store three years' worth of rainfall, removing the threat of drought and thus ensuring a constant supply of water for domestic requirements and irrigation (1), meaning more food available for consumption and for sale which improves the health of the people and the economy of the country (1). Cheap supply of HEP encourages industry into the area, creating jobs and improving the standard of living in the area (1). The reservoirs encourage tourists into the area, improving the local economy, eg the Missouri reservoirs contribute around $100 million to the regional economy each year (1). Improved facilities like boat ramps and campgrounds were built, improving leisure facilities for locals as well as tourists (1). The reservoirs encourage new wildlife into the area such as waterfowl, a new water bird habitat and spawning areas for fish (1).	5
2. Development and Health			
		Primary health care (PHC) strategies have been introduced by many developing countries in an effort to improve the health of the population. In rural areas of China, the 'barefoot doctors' programme has been introduced. Barefoot doctors are local people who are given basic training so they can attend to the health needs of their community, eg snake bites, improving general health (1). This means that medical aid is within easy reach of the community and this saves them travelling to the neareast hospital (1). It also takes the pressure off the large hospitals, allowing them to deal with more serious illnesses (1). The barefoot doctors are effective as they provide the hope of healthcare for people in more remote areas where budgets and manpower can be limited (1). Training costs are low, eg in India it costs $100 to train a health worker for a year (1). Although the barefoot doctors cannot carry out major procedures, they provide people with important information and services such as advice on birth control, vaccinations and basic hygiene which helps prevent the spread of disease, reduce the birth rates and infant mortality rates. This means that the government can spend more money on development (2). In some areas, doctors visit rural villages and treat the sick who are too ill to travel to the nearest hospital. They run clinics in the larger villages and take a mobile health van to the more remote villages. This means that the villagers receive vital healthcare which can reduce the death rate (1). Some PHC programmes provide villages with local dispensaries which improves the health of the people by giving them access to essential modern drugs and family planning (1). Some programmes also implement improvements to sanitation facilities and the provision of clean drinking water, reducing the cases of cholera and malaria and improving the health of the working population which allows them to provide for their families (1).	10
3. Global Climate Change			
	(a)	Extracting and burning fossil fuels such as coal or petroleum results in the release of carbon dioxide (CO_2) and other heat-trapping 'greenhouse gases' into the atmosphere (1). Growing populations and the consequent rise in the use of electrical gadgets increases the demand for electricity, thus increasing the amount of carbon dioxide released into the atmosphere (1). The number of private cars on the road, as well as an increase in the use of lorries to transport goods, eg food shopping online, has increased exhaust emissions entering the atmosphere (1). No-frills airlines like easyJet and Ryanair have made air travel much more accessible to large numbers, and a growing demand for products from all over the world has increased the use of aircraft, and thus the consumption of fossil fuels (1). Clearing forests releases large amounts of CO_2. Also, plants and trees use CO_2 to grow so deforestation means there are less trees to absorb the extra CO_2, meaning more CO_2 stays in the atmosphere, trapping more heat (2). The continually increasing world population means there is a greater demand for food, resulting in more cattle being farmed. Cows produce harmful gases such as methane which contribute to global warming (1).	5
	(b)	Global warming is causing glaciers to melt, putting millions of people at risk from floods, eg the Fenlands of Eastern England and the Ganges Delta in Bangladesh (1). Many low-lying island nations are at risk of submergence from rising sea levels, eg the Kiribati Islands (1), and saltwater intrusion affects the quality of water in wells, and floods taro patches and gardens affecting food supply (1). Warmer sea temperatures can affect sea life, eg Scotland's hottest year on record was in 2003 and this rise in temperature killed hundreds of adult salmon as rivers became too warm for them to extract enough oxygen from the water (1). Warming sea temperatures force fish shoals to move to cooler waters, affecting catches and fishermen's livelihoods (1). In the Sahel area of North Africa, rising temperatures may result in more droughts which means crops won't grow and famines will become more frequent (1). In other areas like the UK, increased temperatures can result in some crops like potatoes failing but at the same time can encourage the growth of soft fruits that like warmer conditions (1).	5

Question	Specific Marking Instructions for this question	Max mark
4. Trade, Aid and Geopolitics		
	The aid may be subject to conditions imposed on the receiving country which ties them to the donor country, eg the receiving country is often required to use the aid to purchase goods and services from the donor country (1). Producing the goods improves the economy of the donor country but might mean that the receiving country could have bought the goods cheaper elsewhere (1). Developed countries may set up industry in the receiving country. Although these industries may provide employment, the profits from the goods produced may go back to the donor country instead of being spent on the necessary needs of the developing country (1). The goods produced may be cheap and the profits low, so the receiving country may not be able to invest in areas such as industry and agriculture (1). In Uganda only 18% of the contract value of World Bank-funded projects went to local firms. Firms from China and the UK won the bulk of large World Bank-funded contracts in Uganda, 32% and 19% respectively (1). Big development banks, including the World Bank, opt for international competitive bidding, increasing the chances that large firms from donor countries will win contracts (1). Two-thirds of formally untied aid contracts still go to firms from rich donor countries, while developing countries are squeezed out by powerful transnational companies (1). Aid may be given to benefit the donor country politically, or to strengthen a military ally, or to provide infrastructure needed by the donor for resource extraction from the recipient country to increase their profit (1). Donor countries give aid to make them look good (1). Companies from developing countries do not have the business knowledge, ability or contacts, therefore lose out on contracts to transnational companies who have business acumen and offices in key areas for the global aid industry such as Washington and Brussels (1).	10
5. Energy		
	Marks could be lost if answer does not include specific examples. Hydro-electric energy involves generating electricity using the power of moving water, therefore, hydro-electric plants need to be located in areas where there is a high average rainfall so that the reservoirs needed to store the water are always full (1). Electricity is generated by the force of the water turning turbines, so steep slopes, eg in mountain areas, are needed (1). Glaciated areas in Scotland with hanging valleys are suitable as they provide the drop necessary to turn the turbines to provide electricity (1). The rock needs to be impermeable to prevent loss of water through seepage (1). In the UK, however, most suitable sites are already used for hydro-electric power stations, so there is very little scope to increase the use of HEP in the future (1). Solar power is suitable in areas where there are long hours of sunshine to power solar panels. This would be more suitable in areas like Mallorca than in Scotland where sunshine hours are more limited (1). Geothermal power is suitable in areas of volcanic activity like Iceland where the power from volcanoes is used to heat houses due to its location on a plate boundary (1). Recently Japan, which is one of world's most seismically active nations, has been investing lots of capital in geothermal power and generates as much as 23 million kilowatts of energy which makes it less dependent on other nations for power (1). Wind power is most efficient in areas with no barriers to the force of the wind, ensuring a continual supply of power to the turbines (1). However, wind power can be controversial as it causes visual pollution as well as disturbing local habitats (1). Wind power is dependent on the continual supply of wind so there is a difficulty in supplying continual energy on calm days as it's difficult to store the energy produced (1). The generation of electricity from wave power is currently under development and, since wave energy resource is distributed across the globe, wave energy offers many countries the benefit of security of supply (1). Bio fuels can provide continuous energy as required by the burning of plant matter and it is affordable in developing countries (1).	10

Section 4: Application of Geographical Skills

Question	Specific Marking Instructions for this question	Max mark
1.	**Sample Answer** There would be a limited impact on the environment as the area is already industrial as the aerial photograph (diagram 2) shows. The new nuclear facility would be built alongside the present Sellafield so would blend in. There are few villages in the area so will not affect very many people. However there are many farms in the area for example Greenmoor Farm at 022053 and that would be lost as the new facility would be built on its site. This would result in less farmland to produce food as well as the farmer losing his livelihood. There is forestry at 026055 and hedgerows in the area which might be removed causing a loss of animal habitat (shown on diagram 2). Also since Moore House Farm is north east of the proposed site then the prevailing wind could blow particles over the farm contaminating his crops and animals therefore not in keeping with the brief for this project. The area is close to a National Park boundary where the landscape is supposed to be protected. The new facility could affect the quality of the park's environment. Local people are up in arms about the risks attached to the present decommissioned power station. The brief says it should have a minimal impact on the people but as the Diagram 4 shows this is not the case as some health problem seems to be linked to the possible effects of radioactivity from the plant. The local newspaper says that radioactive particles have been found on the surrounding beaches causing concerns for the health of the locals as well as polluting the environment. However diagram 3 refutes these claims as Sellafield has won eight safety awards suggesting there is no problem with the risk of radioactivity affecting people and the environment. The brief calls for good transport routes. The area has relatively good access with a train station a few kilometres away at Seascale at grid reference 037011. There is also good road access via the A595 with a minor road connected directly to the power station. Although there are no large towns close by workers can be found in the surrounding villages of Egremont, Gosforth and Seascale. It can be seen from the OS map that the contour lines on the map are quite far apart showing the land is relatively flat. This meets the brief as this flat land will be easy to build on. The area suggested is relatively empty so there is room to expand if necessary. In my opinion there are more disadvantages than advantages. This site does meet the conditions of the brief as far suitability of flat land to build on, access into and out of the area via road and rail and the availability of a nearby workforce. However I think the effect on the local people, environment and landscape with loss of habitat, visual and air pollution and the possible effects on health outweighs the advantages.	10

HIGHER FOR CfE GEOGRAPHY
MODEL PAPER 2

Section 1: Physical Environments

Question	Specific Marking Instructions for this question	Max mark
1.	**For full marks reference must be made to two factors.** The size of the drainage basin has an effect as large basins will have high peak discharges because they catch more precipitation (1). Larger basins have longer lag times than small basins because the water takes longer to reach the rivers (1). Basins with steep slopes will have a high peak discharge and a short lag time because the water can travel faster downhill (1). Permeable rocks and soils, eg sandy soils, absorb water easily allowing water to travel slowly through the soil, reducing peak discharge as well as increasing the lag time in a river (1). Impermeable soils, eg clay soils, are more closely packed so water can't infiltrate, therefore, water reaches the river more quickly reducing the lag time (1). Vegetation intercepts precipitation and slows the movement of water into river channels which increases lag time (1). Water is also lost due to evaporation and transpiration from the vegetation and this reduces the peak discharge of a river (1).	4
2.	**If Brown Earth chosen:** Brown earths have a deep, thick humus layer formed from decaying leaves from deciduous trees like oak when they lose their leaves in autumn (1). Brown earths are formed under warm climates. This encourages the organic material to decay well and adds humus to the soil (1). The horizons' boundaries are not distinct because biota like earthworms mix the soil well and keep it aerated (1). This mixing of the soil by earthworms and the long roots of the plants recycling up the minerals from far below the surface also keeps it fertile (1). The movement up through the soil by capillary action encourages the soil to mix (1). The soil is slightly leached as, in most areas, precipitation is greater than evapotranspiration, resulting in the minerals being washed downwards (1). This causes an iron pan and this hinders drainage leading to flooding (1).	6

Question		Specific Marking Instructions for this question	Max mark
2.		**(continued)** **If Podzol chosen:** Podzols tend to be found on the upper slopes of upland areas where precipitation is heavy or where the vegetation is coniferous forest, producing an acid humus (1). Snow melt is created in mountain areas in spring and this along with the high level of precipitation leads to leaching of minerals creating an iron pan (1). Drainage is poor as the iron pan restricts the passage of water through the soil (1). The climate also affects the colour of a podzol soil as the leaching causes podzolization which removes iron and other minerals creating the ash grey layer (1). Podzols are associated with coniferous forests and the pine needles result in the creation of an acid soil (1). The acid conditions are a deterrent to soil organisms like worms so the soil is not well mixed so distinctive horizons are formed (1). The slow rate of weathering of the parent rock gives a shallow soil which, due to its acidity and lack of humus, is usually infertile (1).	
3.		**A series of fully annotated diagrams could achieve full marks.** Corries form when snow collects in hollows on a mountainside, especially in shaded areas. The snow turns to ice or névé and begins to move downhill as a result of gravity (1). Freeze-thaw weathering, where water gets into cracks in the rock and freezes at night, puts pressure on the rock which eventually cracks (1) and plucking, where ice sticks to the rock and pulls fragments away as it moves, steepens the back wall (1). Material that falls into the glacier causes abrasion on the floor of the hollow as the material acts like sandpaper smoothing and deepening the hollow to form a deep rock basin (1). As the glacier moves out of the hollow, it loses some energy and material is deposited forming a rock lip (1). As the ice retreats and melts, a deep armchair-shaped hollow is left by the plucking and abrasion of the glacier (1).	5

Section 2: Human Environments

Question		Specific Marking Instructions for this question	Max mark
1.		**In this question, you can give positive impacts, negative impacts or both.** Impact on the UK: **Positive impact** Polish migrants are prepared to do menial, unskilled, low-paid jobs in factories, catering, cleaning, and labouring which many UK citizens do not want to do (1). Polish workers are prepared to work long, unsociable hours and for less money than a British worker (1). Job vacancies in the UK and the skills gap can be filled by skilled Polish migrants, eg managerial and professional vacancies (1). The Polish migrants pay National Insurance contributions and these have helped the UK cope with supporting its aging population (1). Migrants bring energy and new ideas, and the culture of the UK is enhanced by the migrants with new foods, fashion and festivals (1). **Negative impact** Tensions can occur with the local population in relation to housing, benefits and healthcare, eg in Boston, Lincolnshire protests have been made to the local council (1). Workers are often exploited by unscrupulous employers with gang masters supplying a labour force for much less than the minimum wage (1). Many immigrants cannot find a job so claim benefits, putting pressure on the finances of the UK government as well as pressure on local authorities to supply education and healthcare for additional people (1).	5
2.		Education can be used to teach people about the importance of the environment, the global and local effects of deforestation, and how they can help save rainforests (1). Save The Rainforest, Inc. offers educational travel tours and student trips to tropical rainforest destinations in Belize, Costa Rica, Galapagos and Panama (1). Afforestation and re-afforestion programmes like the World Bank Amazon Region Protected Programme, where trees are planted in new areas or replanted in felled areas, aim to conserve soils or protect existing forest (1). Companies that operate in ways that minimise damage to the environment are encouraged. They might agree to replace the trees as well as financially helping the region (1). National parks and reserves are areas created by the government and protected by law. These laws prevent or reduce the harm caused to the area by commercial developments (1). **Eco-tourism** is popular within these national parks and reserves and the money generated by tourism is reinvested in the area (1).	6

Question		Specific Marking Instructions for this question	Max mark
3.		**If Rocinha Shanty Town, Brazil chosen:** Self-help housing schemes have been implemented, where residents are provided with the materials and given training and tools to do the job to improve their houses, eg wooden shacks are replaced with brick and tile buildings (1). The local authority is committed to spending £200 million pounds to replace wooden buildings or those on dangerously steep slopes with newer, larger houses which will reduce overcrowding and the risk of lives lost from landslides (1). Electricity along with mains water has been introduced; this improves living conditions and health by allowing access to cookers, fridges to keep food fresh and mains water, reducing the spread of disease from dirty well water (1). Roads have been paved, allowing health workers and goods to be transported through the shanty town (1), and streets have been widened to allow access to emergency vehicles and waste collection vehicles reducing the spread of disease (1). The government is looking at rural investment which involves improving the quality of life in rural areas to encourage people to stay instead of migrating to urban areas and increasing the spread of shanty towns (1).	4

Section 3: Global Issues

Question		Specific Marking Instructions for this question	Max mark
1. River Basin Management			
	(a)	**Physical factors might include:** The rock type must be strong and hard to withstand the weight of the water in the reservoir behind the dam (1). The rock should also be impermeable so that no water is lost through seepage (1). There must be a large catchment area (large river basin) to maintain a high water level in the reservoir; so many streams should flow into the reservoir to provide this (1). The valley should be narrow to reduce the cost of construction because RBM schemes are expensive (1). The total annual rainfall should be high (above 1000 mm) to maintain water levels in the reservoir (1). Temperatures should be as low as possible to reduce the amount of water lost through evaporation (1). Also, high evaporation rates can alter the hydrological cycle, affecting the annual amount of rainfall received thus reducing the storage capacity (1).	5
	(b)	**For full marks you should refer to all three factors.** **If the Colorado River chosen:** There is an improved water supply for drinking as the dams ensure there is a water supply available all year round. This availability of water can support the increasing population, especially in the desert cities of Las Vegas and Phoenix (1). Tourists are attracted to Lake Mead and tourist facilities such as camp sites, water sports and restaurants, not only provide jobs for the local people but improve their leisure pursuits as well (1). The creation of reservoirs like Lake Mead has encouraged new wild life habitats to develop, with more than 250 species of birds being seen in the area (1). The availability of water for irrigation in areas of California and Arizona means that farmers can produce more crops to supply the growing population, develop agribusiness and sell any surpluses to produce income for the economy (1). The availability of cheap electricity produced by the dams encourages industry, improving employment in the area, as well as the standard of living of local people (1).	5
2. Development and Health			
		If Malaria chosen: **This question can be answered as part a (efforts to combat disease) and part b (effectiveness of these efforts) or as a combined answer as shown below. Efforts to combat disease = C. Effectiveness = E.** Drugs like Chloroquine can be given to people to stop them catching malaria or to treat malaria sufferers. Malarone is a newer drug and is safe for children (C1). Chloroquine is cheap to produce, but in recent years the mosquitoes have become immune to it so it is less effective. It needs to be taken properly to be effective and can be too expensive for poor people in countries like Malawi (E1). Malarone has proved to be effective and has no side effects, so it has the potential to do well (E1). Vaccines have been developed but they are still at the trial stage in places like Columbia and Gambia (C1). Vaccines can be administered through primary health care (PHC) schemes and, if given to children, would reduce the incidence of malaria and reduce infant mortality rates in many areas (C1). Educating people is an effective, simple and cheap way of combating malaria and can be done through PHC using songs/drama. Advice includes using bed nets, covering the skin to prevent bites and using screen doors and windows to stop mosquitoes entering the house (C1). Bed nets are relatively cheap and stop people getting bitten, reducing the chance of catching malaria (E1). Visitors and locals can also take anti-malarial drugs, but they are expensive and have to be taken regularly to be effective (E1). Draining areas of stagnant water like puddles and swamps reduces areas where the mosquitoes breed. However, there can be heavy rainfall every day in tropical areas so new areas of stagnant water appear all the time and constantly having to drain the area is time consuming and would need to be done every day (C/E1). Breeding areas and homes can be sprayed with insecticides like DDT which kills the mosquitoes and their larvae.	10

Question		Specific Marking Instructions for this question	Max mark
2.		**(continued)**	
		This is effective if done carefully but over time the mosquitoes become immune and DDT is harmful to the environment and is now banned in some countries (C/E1). Egg whites can be sprayed on the stagnant water to suffocate the larvae by clogging up their breathing tubes so that they drown. This works but, in areas where food is scarce, it is a waste of a valuable food source and only the whites are used so the yokes can be wasted (C/E1).	
3. Global Climate Change			
	(a)	People can reduce the use of aerosol sprays which include CFCs by using roll on deodorants, decreasing the amount of harmful gases going into the atmosphere (1). In the UK, there are strict laws for the disposal of fridges etc which contain CFC gases to ensure they are disposed of safely instead of being dumped on landfill where the gases can escape (1). People can reduce the use of fossil fuels such as coal, oil and natural gases by introducing green-friendly fuels such as hydro-electric power, wind power, solar power and other renewable enrgy sources (1). Laws can be introduced to reduce the burning of the rainforest. Trees store carbon so, when burnt, carbon dioxide (CO_2) is released into the atmosphere (1). Replanting schemes can be introduced where forests have been detroyed to ensure CO_2 is recycled by the forest, reducing the build up of greenhouse gasses (1). People can be encouraged to walk, car share, use public transpot and cycle to work to reduce the amount of vehicles on the road, thus reducing the amount of exhaust fumes building up in the atmosphere (1). People can use lead-free petrol, as well as modified engines with catalytic converters, to reduce pollution into the atmosphere (1).	6
	(b)	In the UK, the introducution of green renewable fuels to replace fossil fuels has been successful as the 15% target set by the government to be acieved by 2020 is on track. However, the planned use of offshore windfarms has proved too expensive and may be abandoned (1). Public transport has been improved to encourage people to leave cars at home. This is partly successful as more people now use public transport, but some people still prefer to have the freedom to travel by car (1).	4
		Companies are working to produce new technology for fuel-efficient engines with reduced emmissions. This has been successful as eco cars are now on the road with engines that shut down when stationary, thus reducing the CO_2 emmissions (1). Many developed countries still do not pay developing countries a reasonable sum for their goods, so countries like Brazil still continue to cut down the rainforest (1).	
4. Trade, Aid and Geopolitics			
	(a)	Developing countries often sell primary products at low value, therefore profits are limited (1). Many countries are producing the same raw material, eg coffee is the staple crop of Brazil, Columbia and Vietnam, which keeps prices low (1). Developing countries are prevented from setting up coffee processing plants by develped countries as they put high import taxes on the processed coffee to protect their own coffee industry (1). Limits and quotas are also put on goods so developing countries find it difficult to compete (1). Developing countries are dependent on one or two raw materials and are vulnerable to price fluctuations (1). Developed countries set the prices for raw materials through trading on commodity exchanges around the world, so developing countries have little say in the sale of their products (1).	5
	(b)	The World Trade Organisation settles trade disputes and promotes free trade and the removal of tarrifs and quotas which can improve the trade of a developing country (1). The removal of trade barriers means that developing countries will have access to lucrative markets in the developed worlds, thus improving their balance of trade (1). Some countries in the Caribbean are attempting to diversify their trade by developing non-traditional exports such as new crops or manufactured goods and are pursuing new markets (1). The creation of trade alliances like the Caribbean Community and Common Market (CARICOM) were established to promote trade between the Caribbean countries (1). Custom duties between member states were removed so even the smallest Caribbean countries have access to a regional market (1).	5
		Member countries are encouraged to purchase raw materials from other CARICOM countries. This has spread the benefits of industrialisation and has encouraged industries to locate in smaller countries (1). The Origanisation of Eastern Caribbean States (OECS) has been established and has created a single currency which makes trade between OECS countries much easier as money is not lost in the transactions (1). The OECS developed a common agricultural policy to subsidise farmers and removed controls on the movement of workers, allowing skilled workers to migrate (1). The Fairtrade organisation was set up and this guarantees a fair price for produce which always covers the cost of production of crops etc regardless of the market price (1). Five-year rolling contracts can be given which allows long-term planning to take place for investment in farm machinery, education etc (1).	

Question		Specific Marking Instructions for this question	Max mark
5. Energy			
		For full marks reference should be made to both developed and developing countries.	10
		The world's population is growing constantly, and is predicted to increase by 25% in the next 20 years. This will increase the demand for energy in both developed and developing countries (1). The majority of the increase will be in developing countries like China and India, as they require energy to continue to develop their economies (1). Rising energy demands from economic output and improved standards of living will put added pressure on the energy supply (1). The high standards of living in the developed countries are attributable to high energy consumption levels, with the use of expensive consumer goods and expansion of IT (1). In developing countries, an increase in standards of living means an increase in the demand for energy to power televisions, washing machines, fridge freezers, air conditioning etc (1).	
		Countries like China are developing manufacturing industries which use up enormous amounts of energy (1). Developing countries increasingly trade bulky primary products with developed countries and these goods need to be transported around the world using up more fossil fuels (1). In countries like India, car ownership is becoming more common due to an increased standard of living, so more energy is needed to run these vehicles (1). Energy increase is less in developed countries because, although the population is growing, it is growing much more slowly and, in cases like Germany, not at all as the death rate is higher than the birth rate (1). Developed countries are being encouraged to conserve energy to reduce the effects of global warming, so energy consumption is growing at a much slower rate (1).	

HIGHER FOR CfE GEOGRAPHY
MODEL PAPER 3

Section 1: Physical Environments

Question		Specific Marking Instructions for this question	Max mark
1.		There is a big contrast in the amount of rainfall between the very dry north area of Gao with only 200mm in the hot desert climate of Mali and the much wetter south of Abidjan with 1700mm in the tropical rainforest climate of the Ivory Coast (1). Bobo-Dioulasso, in Burkina Faso in Central West Africa, has an in-between amount of both rain days and total precipitation, with 1000mm in a Savannah climate. Gao has a limited amount of precipitation in summer, Bobo-Dioulasso has a clear wet season/ dry season regime and Abidjan has a twin peak regime with a major peak in June and a smaller peak in October/November (1). Abidjan, on the Gulf of Guinea coast, is influenced by hot, humid tropical maritime air for most of the year, accounting for its higher total annual precipitation and greater number of rain days (1). The twin precipitation peaks can be attributed to the ITCZ moving northwards in the early part of the year and then southwards later in the year in line with the thermal equator/ overhead sun (1). Gao, on the other hand, is under the influence of hot, dry tropical continental air for most of the year and, therefore, has far fewer rain days and a very low annual precipitation figure as it lies well to the north of the ITCZ for most of the year (1). Bobo-Dioulasso again is in an in-between position, getting more rain days and heavy summer precipitation from June–August when the ITCZ is furthest north (1).	6
2.		**A sequence of diagrams, fully annotated, could score full marks. Answers which fail to make use of a diagram(s) would score a maximum of four.** **If Terminal Moraine chosen:** Moraine is material carried along by the glacier. When the glacier reaches lower altitudes or temperatures rise, the ice melts and deposits moraine at its snout (1). Terminal moraine marks the furthest extent of the glacier. It forms a jumbled mass of unsorted materials that stretches across the valley floor (1). Once the ice has retreated, the terminal moraine can often form a natural dam, creating a ribbon lake (1).	5

Question		Specific Marking Instructions for this question	Max mark
2.		**(continued)** **If Drumlin chosen:** Drumlins are elongated hills of glacial deposits and are formed when the ice is still moving (1). The steep slope faces upstream and the lee slope is the gentler, longer axis of the drumlin which indicates the direction in which the glacier was moving (1). The drumlin would have been deposited when the glacier became overloaded with sediment, and as the glacier lost power due to it melting the material was then deposited (1). It is possible that there was a small obstacle in the ground which caused the till to build up around it forming the drumlin (1). It may also have been reshaped by further ice movements after it was deposited (1). **If Esker chosen:** Eskers are mounds of sand and gravel that commonly snake their way across the landscape and are produced as a result of running water in, on or under the glacier (1). When the glacier retreats, the sediment that had been deposited in the channel is lowered to the land surface where it forms a mound, or hill, that is roughly parallel to the path of the original glacial river (1). The water is unable to escape sideways because of the restricting walls of ice (1) and the stream bed is gradually built up above the level of the ground on which the glacier rests (1). The ice that formed the sides and roof of the tunnel later disappears, leaving behind sand and gravel deposits in ridges with long and twisting shapes (1). **Section Across Glacier** meltwater flowing into crevasses in glacier surface maximum height of water in flooded crevasses ice rock meltwater deposits in ice tunnel **Section Across Glacial Valley** (ice now melted) rock collapsed ridge of meltwater deposits (esker)	
3.		Two factors should be mentioned for full marks. Brown earth soils are rich in soil organisms like earthworm and they break down the leaf debris, producing slightly acidic mull humus. These organisms also mix the soil together, aerating it and preventing the formation of distinct layers in the soil (1). The type of vegetation creates the acidity or alkalinity of the soil and the leaves from the deciduous trees provide a deep leaf litter which is broken down rapidly in the mild climate (1). Tree roots penetrate deep down into the soil ensuring the minerals are recycled back to the vegetation (1). There is slightly more precipitation than evaporation and this will result in a leaching of the minerals which may cause an iron pan (1). These soils are well drained; there is little accumulation of excess water so little leaching occurs (1). Rock type determines the rate of weathering, with hard rocks such as granites taking longer to weather, producing thinner soils, whilst softer rocks such as shale weather more quickly (1).	4

Section 2: Human Environments

Question		Specific Marking Instructions for this question	Max mark
1.		There will be an increase in the economically active population, generating more taxation money for the government (1). The population is now living longer and is healthier, providing a fitter and more productive workforce (1). The population is rising which will increase demand for and put pressure on public services like schools, hospitals and sanitation (1). Pressure on resources like land, building materials and fuel will also grow, leading to an increase in deforestation and pollution (1). A growing population leads to an increased demand for food which may lead to food shortages and dependency on foreign aid (1). Food shortages and lack of provision of public services could lead to malnutrition and the spread of disease (1). Lack of jobs in the countryside encourages rural–urban migration which leads to the growth of more shanty towns (1).	5

Question		Specific Marking Instructions for this question	Max mark
2.		For full marks both people and the environment should be mentioned.	5
		If the Sahel chosen:	
		Over cropping and the cultivation of only one crop (monoculture) leads to nutrients being removed from the soil, resulting in crop failure. There has been a 25% reduction in agricultural production in the Sahel since 2010 (1). This has led to people being under-nourished with many dying from starvation. This has led to famine in Mauritania and Chad (1). When people are under-nourished, they are susceptible to diseases such as kwashiorkor; if infected, they cannot work so have no money to buy food (1). When the crops fail, people in rural areas are forced to leave the countryside and move to the cities in search of food and employment which results in the growth of shanty towns (1). The traditional way life of the nomads is threatened as they cannot find food and water for their animals. Many are forced to settle in villages or at oases. This in turn puts pressure on the surrounding land, leading to over-cultivation (1). Many people become refugees as they are forced to leave their homes and seek food and shelter in neighbouring countries. This can lead to conflict with the resident populations (1).	
3.		**If Rocinha chosen:**	5
		In some areas, housing has improved, with better sanitation and electricity, which has resulted in better living conditions with improved health (1). Some schools have been set up, so some children now get an education with the chance of a better job to improve their standard of living (1). Housing is less crowded, reducing the spread of disease such as diarrhoea (1). However, crime rates are still high, with violence and drug-related crime causing major problems as locals do not trust the police (1). Very few formal jobs are available, so unemployment is still at very high levels amongst its residents (1).	

Section 3: Global Issues

Question		Specific Marking Instructions for this question	Max mark
1. River Basin Management			
	(a)	For full marks, all three sites should be mentioned.	5
		The answer scheme gives detailed information on each site but some points are reversals, eg impermeable rock features are the reverse of permeable rock features so mark would only be awarded either to Site A or to Site B but not both.	
		Site A has impermeable rock, so rain water will not soak into the ground but will run into the rivers and streams increasing the volume of water available to be stored (1). The drainage basin has many tributaries which run into the river so a large volume of water can be carried in the river (1). It lies further upstream from the urbanised area so will not cause visual pollution and would control the flow of water before it reaches the urbanised area (1).	
		Site B has permeable rock, so water could seep away and reduce the amount of water for storage (1). Impermeable rock is stronger to support the weight of the dam (1). Forestry would remove some of the available precipitation through interception, reducing the amount of water available for storage (1). Mountainous areas receive more rainfall and snowmelt in the spring which can keep the reservoirs full (1).	
		Site C has a very small catchment area with few tributaries, so less water would be available for storage (1). It has impermeable rock, so it's suitable to take the weight of the dam as well as reducing seepage (1). It is close to an urban area, so the dam will control flooding and ensure a water supply/power (1).	
	(b)	For full marks both social and environmental disadvantages should be mentioned.	5
		If the Nile Basin chosen:	
		Social disadvantages could include:	
		People have to be moved off their land when reservoirs are created, losing their homes and their livelihoods, eg 90,000 Nubians were forced to move to create reservoirs in the Nile Valley Scheme (1). Static water can cause an increase in water-born diseases such as Bilharzia which is spread by the Bilharzia snail (1).	
		Environmental disadvantages could include:	
		Water in the river and on farmland becomes saline with high evaporation rates, so farmers downstream have to switch to more salt-tolerant crops (1). Poor irrigation techniques have led to the waterlogging of soils, resulting in fewer crops being produced (1). A change in river system has caused the loss of many animal habitats such as wading birds and aquatic mammals (1). The flooding behind the dam has flooded many historical sites such as the Abu Simbel temple, resulting in a loss of heritage for the local people (1).	

Question	Specific Marking Instructions for this question	Max mark
2. Development and Health		
	If PQLI chosen: **Answer to part (a) and (b) are combined here. If you prefer, when you are answering this type of question, you may separate out the explanation and the evaluation.** The PQLI uses more than one development indicator, eg life expectancy, adult literacy rates and infant mortality rates, and therefore gives a more balanced view of the level of development in a country (1). It is measured on a scale of 0 to 100, so the higher the number, the better developed a country is. If a country's PQLI is below 77, the country is said to be poorly developed (1). It can reflect the standard of living in a country, eg the average life expectancy in the UK is 79, whereas in Sierra Leone it is 44 (1). This suggests that the UK is a wealthy country and can afford to look after its people with good healthcare and education (1). The percentage of people with access to clean water is almost 100% in the UK but is much lower in Sierra Leone. A lack of clean water leads to disease so people become ill and cannot work; this is reflected in its PQLI rating (1). The UK has well-paid jobs in secondary, tertiary and quaternary industries, producing money for the country to develop, whereas in Sierra Leone most people are involved in subsistence agriculture so find it difficult to feed themselves, and the government is left with little money to invest in healthcare, education etc which all lead to development (1). Where literacy rates are high, 99% in the UK as opposed to 35% in Sierra Leone, it means that the country has money to spend on schools and training teachers which results in an educated workforce which, in turn, produces wealth that the government can use for development (1). Problems exist when trying to evaluate the development of a country using only social indicators like literacy rates; they are only averages across a country and therefore hide regional differences, and they also don't compare health with education and diet (1). Individual indicators like GNP are based on an average and so may be easily skewed by a few very wealthy families which may mask extreme poverty for the majority of the population (1). A single economic indicator is inaccurate as it does not show the wealth of a country and it doesn't show data on how well educated people are or how good their diet is (1). It does not take into account differences between urban and rural areas or, e.g. in Brazil, between poor favelas and richer inner cities (1). When evaluating the usefulness of the PQLI, clearly this measure provides a much more accurate assessment of the level of development of any given country. It also allows for comparisons of development levels between countries especially in terms of social and economic development (1).	10
3. Global Climate Change		
(a)	**Human activities:** Vehicles use a tremendous amount of fossil fuels like petrol and oil. These fossil fuels have been storing carbon for thousands of years but when used in vehicles a large amount of carbon dioxide (CO_2) is released into the atmosphere (1). Increasing population numbers result in land being cleared for housing, industry and HEP. This leads to trees being destroyed and when trees are removed they stop storing carbon and release the carbon they have accumulated over their lifetime (1). Eating supermarket meat, which requires both the clearing of land for raising animals and growing feed crops, eg cattle ranching in the rain forest and the use of machines used in the production, processing and transporting the produce, creates large emissions of CO_2 (1). Also, increasing populations demand more meat, so more cows/animals are kept. Cows produce methane, adding to the greenhouse gasses in the atmosphere (1).	6
(b)	**For full marks both people and the environment should be mentioned.** Increased temperatures could make it too hot to grow certain crops, as well as causing droughts which could reduce the amount of water available for irrigation (1). Climate change is also likely to cause stronger storms and more floods, which can damage crops (1). Higher temperatures and changing rainfall patterns could result in some kinds of weeds and pests spreading to new areas (1). Less snowpack and earlier snowmelt will reduce the amount of water flowing into rivers like the Colorado, and many places rely on snowmelt to fill the lakes, rivers, and streams that help keep drinking water reservoirs full and provide water to irrigate crops (1). The amount of bad ozone is likely to increase as more ozone is created when temperatures are warmer and this will adversely affect people with asthma (1). Climate change might allow some infectious diseases to spread. As winter temperatures increase, ticks and mosquitoes that carry diseases can survive longer throughout the year and expand their ranges, putting more people at risk (1). Climate change poses risks for cities near the ocean. Places like Miami, New York City, New Orleans and Venice could flood more often or more severely if sea levels continue to rise (1). If that happens, many people will lose their homes and businesses (1).	4

Question	Specific Marking Instructions for this question	Max mark
4. Trade, Aid and Geopolitics		
	Governments of developed countries which donate foreign aid have to agree the amount of money which is given to aid projects. Normally, this is a percentage of their gross national product (1). This aid may form part of a bilateral agreement between the donor and the receiving country, or it may be a sum allotted to various international agencies such as the World Health Organisation (1). Some countries insist on a scheme of tied aid, whereby the receiving country must use aid funds to purchase goods and services from the donor country (1). This involves political agreements between donor and receiver, and this kind of aid benefits the donor country more than the receiving country (1).	10
	Short-term aid is given after a natural disaster and a government makes a political decision about the amount. This type of aid does not involve tying the receiving country to an agreement (1). Long-term aid is given over a long period of time and the donor country may retain some control over the receiving country as they can control how the money is spent (1). Voluntary aid may be given to countries through international charities such as Oxfam, the Red Cross and Medecins Sans Frontieres where an international network of volunteers, field projects and offices aids victims of armed conflict, epidemics, and other disasters (1). Political decisions would be taken to avoid the funds of these agencies having to pay tax (1). The internal politics of the receiving country may influence how foreign aid is spent. Some governments may use some of the money to build up armed forces or buy weapons (1). A donor country will look at the political system of the developing country before deciding whether or not to give aid, eg it may decide not to send any aid to a country which is not a democracy (1). The donor country may want to make an ally of the recipient country which may be useful in a future conflict (1). Giving aid to a developing country may give the donor country prestige within the international community. It may also win support at home for the government of the donor country (1).	
5. Energy		
	The distribution of fossil fuels is highly uneven and depends on geological structure and therefore is accessible in specific geographical areas (1). Most gas and oil is found in areas with sedimentary basins and close to plate boundaries, e.g. the Middle East which has 60% of the world's oil reserves and 40% of its gas reserves (1). Some developing countries do not have the technology to develop these resources, so they are inaccessible to them (1). Richer developed countries have the money and expertise to extract these fossil fuels, e.g. the UK has access to North Sea oil (1). Areas like Iceland, due to its geographical location on a plate boundary, make use of geothermal power to heat its houses etc (1). Solar power is used in countries like Spain which have a sunnier climate than countries like the UK and can make use of sunshine to produce solar energy (1).	10
	Hydro-electric power (HEP) is suitable in areas where there is a high rainfall, eg mountainous areas of Scotland and Alaska. Countries with glaciated landscapes are suitable too, as corrie lochs and hanging valleys can be used or dammed to hold more water, and the drop used to turn turbines (1). Nations with coastlines have access to the sea where wave power can be harnessed to produce electricity (1). Countries like Japan with limited natural resources of coal and gas have invested in the production of electricity using Nuclear Power (1). Many poor nations have limited access to fossil fuels so they exploit the natural resources of the area, relying on trees/bushes etc to use for firewood for cooking, lighting etc (1).	

HIGHER FOR CfE GEOGRAPHY
2015

Section 1: Physical Environments

Question		General marking principle for this type of question	Max mark	Specific Marking Instruction for this question
1.		Award 1 mark for a developed explanation, two limited explanations, or for a description of the discharge with limited explanation. Award a maximum of 2 marks for description of the discharge with 1 mark being awarded for every two descriptive points being made. Markers should accept all relevant and appropriate explanations for the source provided.	4	Slight increase in discharge until 09.00 hours in response to rain which started to fall at 07.00 hours (1 mark). At first, this rain would have been intercepted by vegetation and have infiltrated the soil (1 mark). There is a steep rising limb up to a peak discharge of 100 cumecs at 18.00 hours (1 mark). This water would have filled up storages in the soil due to throughflow and groundwater (1 mark) — as the soil became saturated, surface run-off increased causing a peak (1 mark). The rising limb becomes less steep briefly between 13.00 and 15.00 hours, caused by a marked reduction in rain to 4mm around 10.00 hours (1 mark). There is a short lag time of 5 hours which could be due to deforestation/steep slopes/impermeable rock (1 mark). A high number of tributaries may lead to the short lag time as water is transported more rapidly by surface run-off (1 mark).The river discharge quickly decreases, shown by a steep falling/recession limb down to because of no more rainfall after 15.00 hours (1 mark).
2.		Award a maximum of 4 marks for either feature. Award a maximum of 6 marks if no diagrams are used. Check any diagram(s) for relevant points not present in the text and award accordingly. Well-annotated diagrams that explain conditions and processes can gain full marks. Award a maximum of 1 mark for three or more correctly named, but undeveloped, processes. Award a maximum of 2 marks for fully developed processes for any one feature. Answers which are purely descriptive, or have no mention of any processes or conditions, should achieve no more than 2 marks in total, with 1 mark being awarded for every two descriptive points being made. **1 mark** Limited explanation — the use of the names of at least two processes in context with no development of these. **2 marks** The use of the names of at least two processes with development of these, but no other reference to conditions. **OR** Limited use of the names of at least two processes, with at least two descriptive points about the landscape formation.	7	Headland with weaknesses such as joints, faults or cracks is eroded by the sea to form firstly caves (1 mark). Erosion takes place due to hydraulic action — pounding waves compress trapped air in the rocks, creating an explosive blast which weakens and loosens rock fragments (1 mark), abrasion/corrasion — rock fragments thrown against the headland create a sandblasting (abrasive) action, wearing away the rock (1 mark), solution/corrosion — carbonic acid in sea water weathering limestone and chalk (1 mark), attrition — rock fragments slowly being ground down by friction from wave action into smaller and rounder pieces (1 mark). In some cases, a blowhole can form in the roof of the cave as compressed air is pushed upwards by the power of the waves, causing vertical erosion (1 mark). Over time, erosion on both sides of the headland cuts through the backwall and enlarges the cave to create an arch (1 mark). Continued erosion at the foot of the headland and the effects of vibrations on the roof of the arch weakens it, eventually resulting in the collapse of the arch roof, leaving a stack isolated from the headland (1 mark). These low ridges of sand or shingle slowly extend from the shore across a bay or a river estuary and are caused by longshore drift (1 mark). This lateral movement occurs when waves, driven by the prevailing wind, pushes material up the beach; known as the swash (1 mark). The returning backwash is dragged back by gravity down the beach at right angles (1 mark). Material slowly builds up to appear above the water and begins to grow longer and wider. The spit develops as long as the supply of deposits is greater than the amount of erosion (1 mark). Spits form when there is a change in direction on a coastline, which allows a sheltered area for deposition (1 mark). They can also develop at a bay or a river estuary where the river current prevents the spit from extending right across the bay or estuary (1 mark). The shape can change through time to become curved or hooked at the end in response to changes in wind direction and currents (1 mark).

Question	General marking principle for this type of question	Max mark	Specific Marking Instruction for this question
2.	**(continued)** **3 marks** Two developed processes with limited explanation of how the feature forms over time. **4 marks** Two named processes with development of these, with two further statements explaining the formation of the feature.		
3.	Award 1 mark for each developed explanation or for two less developed points. Candidates should be awarded 1 mark if they only name and locate at least two cells/winds, not credited elsewhere. Credit any valid responses.	4	Warm air rises at the Equator and travels in the upper atmosphere to around 30^0 N and S, cools and sinks **(1 mark)**. Air moves from the tropical high to the low pressure area at the equator creating the Hadley cell/ Trade Winds **(1 mark)**. Cold air sinking at the poles moves to 60°N/S to form the Polar Cell/Polar Easterlies **(1 mark)**. The cold air from the poles meets warmer air from the tropics, causing air to rise creating the Ferrel Cell low pressure **(1 mark)**. Air is moved from the tropical HP, towards LP at the polar front, forming the westerlies **(1 mark)**. This convergence causes the air to rise, with some of this flowing in the upper atmosphere to the Poles where it sinks, forming the Polar cell. Easterly winds blow away from the high pressure at the Pole. **(1 mark)**. Warm air from the Equator is distributed to higher and cooler latitudes and cold air from the Poles distributed to lower and warmer latitudes **(1 mark)**. Due to the Coriolis effect winds are deflected to the right in the northern hemisphere **(1 mark)**. Credit should be awarded for answers which refer to the Rossby Waves and/or the Jet Stream.
4.	Candidates should discuss the positive or negative consequences of the predicted population structure in 2050. For 1 mark, candidates may give one detailed consequence, or a limited description/explanation of two factors. Detail may include relevant exemplification of a problem. A maximum of 2 marks should be awarded for answers consisting entirely of limited descriptive/explanatory points, with two such points required for 1 mark. Care should be taken not to ensure consequences are relevant to developing countries. Credit any other valid responses.	5	*Possible answers may include:* • The total population will increase significantly putting additional pressure on services and resources like education **(1 mark)**. Housing in Ghana, like many developing countries, is already overcrowded. This problem is likely to continue, with many people being forced to live in shanty town housing **(1 mark)**. • There will be a much larger potential workforce which may attract multinational companies to the country **(1 mark)**. An increase in the active age group, however, could also result in higher levels of unemployment or underemployment **(1 mark)**. • In total, there will be twice as many children, so significant investment in maternity hospitals, immunisation programmes and education will be needed **(1 mark)**. It will be necessary to build more schools and train more teachers to support the growing number of young people **(1 mark)**. • Government policies may promote smaller families or encourage emigration to reduce the problems of over-population **(1 mark)**. • With life expectancy increasing, it will also be necessary to invest in health-care to meet the needs of an ageing population in the future **(1 mark)**.

Question	General marking principle for this type of question	Max mark	Specific Marking Instruction for this question
5.	Answers will depend on the case study referenced by the candidate. Marks may be awarded as follows: For 1 mark, candidates may give one detailed explanation, or a limited description/explanation of two factors. A maximum of 2 marks should be awarded for answers consisting entirely of limited descriptive points, with two such points required for 1 mark. A maximum of 4 marks should be awarded if the answer does not clearly relate to a specific case study. Credit any other valid responses.	5	*For candidates who write about Glasgow, possible answers might include:* • Bypass/ring road: In Glasgow, the M74, M77 or Glasgow Southern Orbital Route mean that through traffic does not need to travel into the city centre **(1 mark)**. • Pedestrianised area in centre: sections of Argyle Street and Buchanan Street were pedestrianised, the main shopping streets are no longer congested to make it safer and more pleasant for shoppers **(1 mark)**. • Park and ride and improvements to public transport: Commuters have been encouraged to travel into the city centre by train, underground or bus to reduce the number of cars on the road **(1 mark)**. Additional parking has been provided at suburban train stations, eg Hamilton West and Carluke, and new railway stations have opened, eg Larkhall **(1 mark)**. The introduction of bus lanes has reduced journey times along main commuter routes, therefore encouraging people to travel by bus **(1 mark)**. • One way streets: To improve the flow of traffic, the streets around George Square and Hope Street are now one way **(1 mark)**. • Parking restrictions and fines: parking charges have been increased (60p for 12 minutes in the city centre, with a maximum stay of 2 hours) and traffic wardens patrol the streets to discourage people from bringing their cars into the centre (£60 fines are issued as necessary) **(1 mark)**. • Multi-storey car parks: Multi storey car parks have been built, eg Buchanan Galleries, to reduce the number of cars parking on the surrounding streets **(1 mark)**. • Bridges and tunnels: Bridges like the Clyde Arc ('Squinty Bridge') and the Clyde Tunnel have been built to improve the flow of traffic from the north to the south of the city, diverting from the CBD thus reducing congestion **(1 mark)**.
6.	Answers will depend on the case study referenced by the candidate. Marks may be awarded as follows: For 1 mark, candidates should briefly describe a strategy and offer one evaluative point. Further developed/detailed evaluative comments should be awarded 1 mark each. At least two strategies are required for full credit. A maximum of 3 marks should be awarded for answers consisting entirely of limited evaluative points. Up to 2 marks can be awarded for description or explanation of strategies (i.e. no evaluation), with two such points required for 1 mark. Credit any other valid responses.	5	*Possible answers might include:* • **Large scale redevelopment:** The Dharavi Redevelopment Project where local people will be moved to high rise apartment blocks, however, only those who have been resident in Dharavi since 2000 will be eligible to move into these apartments **(1 mark)**. Other residents will be moved to other parts of the city, which will break up communities and may result in people being too far from their work **(1 mark)**. The new flats will also be too small for those who currently have workshops above their homes **(1 mark)**. • **Slum Rehabilitation:** has planned and managed improvements such as upgrading mains sewerage to help reduce diseases such as cholera however, within twelve years, only 15% of Dharavi was redeveloped **(1 mark)**. • **Local projects:** Self-help schemes support the efforts of local people to improve their housing for example by adding an additional floor to buildings thus reducing overcrowding **(1 mark)**. Toilets have been added and are shared by two or three families who help to keep them clean, which has reduced the incidence of water related diseases **(1 mark)**.

Question		General marking principle for this type of question	Max mark	Specific Marking Instruction for this question
7.	(a)	1 mark should be awarded for each detailed explanation. Award a maximum of 4 marks if only human or physical factors are explained (although a physical factor could be linked to a human issue, for example narrow cross section which would reduce the cost) A maximum of 2 marks should be awarded for answers consisting entirely of limited descriptive points, with two such points required for 1 mark. Credit any other valid responses.	5	Narrow cross section of the valley is required in order to reduce construction costs of the dam (1 mark). A deep valley is required behind the dam as this will result in a smaller surface area for the reservoir, thereby reduce loss from evaporation (1 mark). A site which has impermeable rock would be advantageous as this would reduce loss from the reservoir by percolation (1 mark). A site which is close to construction materials would help to reduce the cost of transporting these materials to the construction site (1 mark). An area free from earthquakes or subsidence is needed as the area needs to be able to support the weight of a large dam (1 mark). A site close to areas of farmland or urban areas would help to reduce water/electrical loss during transportation (1 mark). The costs involved in moving people who live in the area to be flooded, to reduce costs for compensation and re-housing (1 mark). The land which is to be flooded should not be valuable, for example high quality farmland or of historic/environmental importance (1 mark). An area with plentiful supply of snowmelt/rainfall/river water is required to ensure a consistently high volume of water in the reservoir (1 mark).
	(b)	Award 1 mark for each detailed explanation. Candidate answers must include both social and economic consequences. No marks should be awarded for negative impacts. Award a maximum of 4 marks if the answer does not clearly relate to a specific named water management project. A maximum of 2 marks should be awarded for answers consisting entirely of limited descriptive points, with two such points required for 1 mark. Credit any other valid responses.	5	**Answers will depend on the water management project chosen but for the Aswan High Dam, possible answers might include:** • Increased access to clean drinking water reduces water borne diseases such as typhoid (1 mark). • Increased irrigation, (with 33,600km^2 of irrigated land) which allows for two crops a year to be grown, reducing malnutrition (1 mark) production of wheat and sugar cane tripled allowing more export crops to be produced (1 mark). • Increase in hydro-electric power (from the 12 generating units in the Dam, these generate approx. 2.1 gigawatts) attracting industries such as smelting industries (1 mark). • The introduction of the Nile perch and tiger fish into Lake Nasser has increased the commercial fishing industry and fishing tourism industry (1 mark). • Industries which require large amounts of water have grown up near to Aswan, for example the Egyptian chemical industry KIMA which makes fertilisers (1 mark).
8.	(a)	1 mark should be awarded for each detailed comparison, or a comparison with a short explanation. A detailed comparison will contain a qualitative statement such as dramatically higher and be supported by the statistics. A maximum of 2 marks should be awarded for answers which are purely descriptive and do not go beyond making comparisons directly from the table, with two such comparisons required for 1 mark. Where candidates refer only to the headings (i.e. not to the data), a maximum of 1 mark should be awarded. Markers should take care to look for comparisons wherever they occur in a candidate's answer. Credit any other valid responses.	4	GDP per person shows that Brazil, Mexico and Cuba are emergent developing countries with an intermediate level of income, while Kenya and Malawi are still classed as low income developing countries (1 mark). Employment in agriculture is far higher in Malawi (90%) than Brazil at 16% so it is less industrialised (1 mark). Adult literacy is far higher in Brazil, Mexico and Cuba than in Malawi suggesting they have more schools (1 mark). Birth rate shows a vast difference in development with both African countries having high birth rates (Kenya – 30%); compared with Mexico at 18%, suggesting lack of contraception/education (1 mark). There is also a large difference between the country with the poorest life – Malawi has a life expectancy of 53 whereas Cuba is 78, suggesting better health care (1 mark).

Question		General marking principle for this type of question	Max mark	Specific Marking Instruction for this question
	(b)	1 mark should be awarded for each detailed explanation, or for two more straightforward explanations A maximum of 2 marks should be awarded for four straightforward descriptive lists exemplified by country name, eg Nigeria can sell oil to make money. There is no need to mention the countries in the question and other countries can be used to exemplify points made. Credit any other valid responses.	6	Some countries have natural resources such as oil, which can be sold to generate foreign currency (1 mark). Some countries eg Chad are landlocked, are find it more expensive to export and import goods (1 mark). Countries with a poor education system have many low skilled workers and are unable to attract foreign investment (1 mark). Countries with fertile soils and a suitable climate can grow cash crops which can be sold for income (1 mark). Corruption in government such as in Nigeria can lead to money being used inappropriately (1 mark). Where countries suffer from conflict or civil war they are unable to keep the economy working and spend extra finance on weapons (1 mark). Countries which have accumulated large debts have to repay loans and interest causing less money for services (1 mark). Famine can lead to malnutrition, and a reduced capacity to work and create income (1 mark).
9.	(a)	No marks should be awarded for naming the greenhouse gases as these are included in the diagram. No marks should be awarded for explanations of the enhanced greenhouse effect as this is included in the diagram. 1 mark should be awarded for each detailed explanation, or for two undeveloped points. Undeveloped points may include comparative potency of each gas, or limited reasons for increase. Markers should take care not to credit the same reason twice, eg car ownership. Credit any other valid responses.	4	*Possible answers might include:* The Enhanced Greenhouse effect has been caused by an increase in greenhouse gases within the atmosphere. **Carbon Dioxide (remains in the atmosphere for 100 years):** • Burning fossil fuels, for example coal, oil and natural gas release carbon dioxide into the atmosphere, which will trap heat. Coal has been used increasingly to power factories, generate electricity in power stations and to heat homes (1 mark). • Increased car ownership has resulted in more petrol and diesel being used to fuel cars (1 mark). • Deforestation, especially in the Amazon Rainforest, has resulted in less carbon dioxide being absorbed, and the burning releases more CO_2 (1 mark). • Peat bog reclamation (particularly in Scotland and Ireland) during, for example, the construction of wind farms, has also resulted in additional carbon dioxide being released into the atmosphere (1 mark). **Methane:** (More than 20 times as effective in trapping heat than CO_2; accounts for 20% of the enhanced greenhouse effect; remains in the atmosphere for 11–12 years) (1 mark). • Methane has been released from landfill sites as waste decomposes and when drilling for natural gas (1 mark). • The increase in padi fields to feed rapidly growing populations in Asian countries has increased the amount of methane in the atmosphere (1 mark). • The increasing demand for beef has resulted in more methane being created by belching cattle and from animal dung (1 mark). **Nitrous oxide:** • Nitrous oxide is 200–300 times more effective in trapping heat than carbon dioxide. Increased car exhaust emissions have resulted in more nitrous oxide (1 mark). • Due to rising food demand the increased production of fertilisers also adds to the amount of nitrous oxide in the atmosphere (1 mark).

Question		General marking principle for this type of question	Max mark	Specific Marking Instruction for this question
9.	(a)	(continued)		**Chlorofluorocarbons (CFCs):** • Refrigerators which are not disposed of correctly release CFCs when the foam insulation inside them is shredded **(1 mark)**. • Air-conditioning — the coolants used in air conditioning systems create CFCs, which must be disposed of correctly **(1 mark)**. Credit any relevant developed points.
	(b)	For 1 mark, candidates should give one detailed explanation, or a limited description/explanation of two factors. A maximum of 3 marks should be awarded for answers consisting entirely of limited descriptive points, with two such points required for 1 mark. A maximum of 2 marks should be awarded for answers consist of limited description with a specific example. 1 mark can be awarded where candidates refer to **two** specific named examples accurately linked with an effect (species, ocean currents, or use of figures). Candidates should be credited for both positive and negative effects. Credit any other valid responses.	6	*Possible answers might include:* **Global Effects:** • Sea level rises caused by an expansion of the sea as it becomes warmer and also by the melting of glaciers and ice caps in Greenland, Antarctica, etc **(1 mark)**. • Low-lying coastal areas, eg Bangladesh affected with large-scale displacement of people and loss of land for farming and destruction of property **(1 mark)**. • More extreme and more variable weather, including floods, droughts, hurricanes, tornadoes becoming more frequent and intense **(1 mark)**. • Globally, an increase in precipitation, particularly in the winter in northern countries such as Scotland, but some areas like the USA Great Plains may experience drier conditions **(1 mark)**. • Increase in extent of tropical/vector borne diseases, eg yellow fever as warmer areas expand, possibly up to 40 million more in Africa being exposed to risk of contracting malaria **(1 mark)**. • Longer growing seasons in many areas in northern Europe for example, increasing food production and range of crops being grown **(1 mark)**. • Impact on wildlife and natural habitats, eg extinction of at least 10% of land species and coral reefs suffer 80% bleaching **(1 mark)**. • Changes to ocean current circulation, eg in the Atlantic the thermohaline circulation starts to lose impact on north–western Europe, resulting in considerably colder winters **(1 mark)**. • Changes in atmospheric patterns linking to changes in the monsoon caused by El Nino, La Nina **(1 mark)**. • Increased risk of forest fires, for example in Australia and California due to change in surface temperatures and changes in rainfall patterns **(1 mark)**.
10.	(a)	Candidates should refer to both social and economic impacts for full credit, although markers should be aware that many impacts could be considered both social and economic. 1 mark should be awarded for each detailed explanation. A maximum of 3 marks should be awarded for answers consisting entirely of limited points, with two such points required for one mark. A developed point may be a detailed explanation of an impact, or may be two less detailed impacts. Candidates may choose to answer a mixture of two styles. Credit any other valid responses.	4	*Possible answers might include:* • Economic impacts include many people being trapped in poverty and trying to survive on very little money. This might be because companies in the developed world want to manufacture their product for the cheapest price possible **(1 mark)**. • Many countries are unable to make a decent profit on the goods they sell because they are forced to pay tariffs to "developed countries" in trading blocs **(1 mark)**. • Government subsidies and grants in "developed countries" allow companies to sell products (eg rice and grain) at a cheaper price than is possible in many "developing countries" — undercutting local farmers and causing them to lose money **(1 mark)**. • Social impacts of this poverty include many children working instead of going to school, and this causes an illiterate population, with the consequent lack of opportunities and poorer quality of living for people that this applies to **(1 mark)**.

Question		General marking principle for this type of question	Max mark	Specific Marking Instruction for this question
10.	(a)	(continued)		• Creates a self-perpetuating cycle of poverty where the next generation are unable to access well paid employment due to being illiterate **(1 mark)**. • The lack of well-paid jobs means many people live in shanty town type accommodation, with little access to clean water, safe electricity, sanitation, etc **(1 mark)**.
	(b)	**Using the information in Diagram Q10b, and also your knowledge of fair trade, explain how fair trade:** **(i) helps to reduce inequalities in world trade, and;** **(ii) impacts on farming communities.** 1 mark should be awarded for each developed point. A maximum of 4 marks should be awarded for answers consisting entirely of limited points, with two such points required for 1 mark. A developed point may be a detailed explanation or a description of a trend with a less detailed explanation or may be two less detailed explanations. Credit any other valid responses.	6	*Possible answers might include:* • The graph shows that the price of normal coffee is very variable — for example when there was a drought in Brazil, the global production decreased and so the price of coffee increased **(1 mark)**. • When China and India increased demand this increased competition for coffee and caused the price to increase **(1 mark)**. • When the production of coffee increased this eased competition for coffee and caused the price to drop **(1 mark)**. • In all these cases, fair trade matched any price increases, but also never dropped below 125 cents per kilo. This allows farmers benefitting from fair trade to have security for the future and to make plans to improve their quality of life **(1 mark)**. • Fair Trade Coffee also does not lose money to the extra "links" in the chain that normal coffee has. Each link in the chain (for example the "middle men" companies), demand a fee and take money away from the farmer. This leaves the farmer with almost nothing to show for growing the coffee in the first place **(1 mark)**. • Fair Trade also ensures that the environment of the coffee plantations is not misused and damaged for future generations by farming in a sustainable way **(1 mark)**. • Fair Trade also ensures that there are no child workers, so that children get the chance to attend school and improve their opportunities in the future **(1 mark)**. • Fair Trade not only benefits the individual farmers, as there are community improvement payments to villages supplying fair trade coffee beans. This allows for example, new schools and health centres to be built and staffed that would otherwise not have been built **(1 mark)**.
11.	(a)	1 mark should be awarded for each developed point for energy production or for two undeveloped points. A developed point may include a descriptive statistic/comparative statement with explanation or a developed explanation. Where candidates have only described the data, award a maximum of 1 mark. 1 mark should be awarded for two descriptive points. Credit any other valid responses.	4	*Possible answers might include:* • Saudi Arabia gets 100% of its energy from fossil fuels because it has vast natural resources of oil and gas in its territory **(1 mark)**. • Portugal has the largest amount of other renewables because its climate is sunny, with many cloud-free days for solar power **(1 mark)** and windy with mid-latitude westerly winds blowing over Atlantic Ocean to Portugal for wind power **(1 mark)**. • France has the largest amount of nuclear energy because it has no oil or gas reserves and has used up most of its coal reserves **(1 mark)**. France used to import oil and gas, but started to develop its own nuclear power in a big way after the "oil shock" — when OPEC countries suddenly decided to quadruple the price for oil in 1973 **(1 mark)**. • Paraguay has the largest amount of Hydroelectric Power because it has large amounts of tropical precipitation and major rivers that can be harnessed **(1 mark)**. • Iceland has the largest amount of geothermal energy because it is located at a plate boundary where the heat of magma is closer to the surface and so can be harnessed efficiently **(1 mark)**.

Question		General marking principle for this type of question	Max mark	Specific Marking Instruction for this question
	(b)	1 mark should be awarded for each developed explanatory point or for two undeveloped points. Take care not to credit reverse statements (for example renewable sources do not pollute and non-renewable sources pollute). Maximum of 4 marks for one energy source. Credit any other valid responses.	6	*Possible answers for all renewable energy sources might include:* • It harnesses a free resource, which is sustainable for the future **(1 mark)**. It does not contribute greenhouse gases such as CO_2 or other pollution in to the atmosphere **(1 mark)**. **Also, for:** • Wind power can be very effective if sited properly in an exposed area where there is regular wind flow to turn the blades **(1 mark)**. A large are of open land is required, close to large cities, for both wind and solar to avoid loss when transporting **(1 mark)**. However, when the wind is not blowing the turbines cannot generate any energy, and at present it is very difficult to store wind power to use when demand needs it **(1 mark)**. Wind turbines built on peat bogs require the bogs to be drained — which releases carbon dioxide that would otherwise have been stored — causing pollution **(1 mark)**. • Solar power is effective on clear, sunny days but is far less effective when clouds block direct sunlight **(1 mark)**. Solar power can be stored in batteries and used at times of high demand **(1 mark)**. Photostatic panels and solar panels have a very high initial cost. **(1 mark)**. • Hydroelectric power can be generated as required and so is able to cope with peaks and troughs in demand **(1 mark)**. • Geothermal power can be generated as required and so is able to cope with peaks and troughs in demand **(1 mark)**. It is most effective in areas close to plate boundaries where the heat source underground (magma) is closer to the surface **(1 mark)**. • Tidal power has the potential to generate vast quantities of energy. Eg experimental tidal generators in the Pentland Firth **(1 mark)**. Tidal power, unlike wind/solar is predictable, therefore reliable. **(1 mark)**. • Technology for wave power, is not yet sufficiently advanced to allow for large scale economic production. **(1 mark)**. *Possible answers for all non-renewable energy sources might include:* It can generate power as required and so is able to cope with peaks and troughs in demand **(1 mark)**. However, it harnesses a resource, which will eventually run out and then will effectively be gone for good, so is not a viable long-term energy solution **(1 mark)**. Coal, gas and oil contribute towards global warming and cause air pollution **(1 mark)**. **Also for:** • Coal/gas are a relatively cheap way for a country to generate energy. They can also be moved about relatively easily to where power is being generated **(1 mark)**. • Gas causes less pollution than burning coal (up to 45% less) or oil (up to 30% less) **(1 mark)**

Question		General marking principle for this type of question	Max mark	Specific Marking Instruction for this question
11.	(b)	(continued)		• Nuclear — due to small amounts of radioactive material needed to produce huge amounts of energy resources will last longer **(1 mark)**. Nuclear power does not produce greenhouse gases such as CO_2 or other forms of air pollution **(1 mark)**. However, it does produce nuclear waste, which is toxic and needs to be carefully stored for hundreds of thousands of years **(1 mark)**. Also, nuclear reactors are expensive to build and run so are not suitable for all countries **(1 mark)**. A radioactive leak could be devastating for the local people and environment and for future generations **(1 mark)**. • Technology in many developed countries is geared towards non-renewable sources as power stations/infrastructure are already in place **(1 mark)**. • Supply of oil may be disrupted and process rise due to sabotage during wars **(1 mark)**.
12.		Candidates should make reference to all sources, including the OS map to discuss the suitability and impact of the centre. 1 mark should be awarded for each developed explanatory point. 1 mark should be awarded where candidates refer to the resource and offer a brief explanation of its significance (beyond the wording of the resource), or give a limited description/explanation of two factors. A maximum of 5 marks should be awarded for answers consisting entirely of limited descriptive points, with two such points required for 1 mark. A maximum of 4 marks should be awarded for candidates who give vague over-generalised answers, which make no reference to the map. There are a variety of ways for candidates to give map evidence including descriptions, grid references and place names. It is possible that some points referred to as a disadvantage may be interpreted by other candidates as a negative impact. Markers should take care to credit each point only once, where it is best explained.	10	*For suitability, possible answers may include the following positive observations:* **Possible advantages of this location** The surrounding area offers potential for a range of activities such as hillwalking, with steep slopes posing a challenge for mountain biking eg Auchenroy Hill **(1 mark)**, there are numerous trails in nearby woodland which could be used for orienteering eg Galloway Forest Park 488038 **(1 mark)**. Steep, sheer cliffs in the disused quarry at 457054 could be used for abseiling/climbing **(1 mark)**. There are a number of opportunities for watersports such as the River Doon could be used for fishing, or the River Doon and Bogton Loch/Loch Doon could be used for kayaking **(1 mark)**. Local schools at 477060 may be interested in using the facilities for outdoor learning and enriching the curriculum. The SSSI at Bogton Loch and the Nature Reserve at Dalmellington Moss would offer opportunities for exploring the wildlife present **(1 mark)**. Variety of attractive scenery including Dalcairnie Linn waterfall offering potential for geography studies (as per spec) at 467043 **(1 mark)**. There are a number of tourist facilities already in the area (Industrial Railway Centre, Horse Riding, Fishing), and these could work together with the new facility to join up and benefit from each other's custom **(1 mark)**. The centre would be only 0.5 km from the A713 main road and there is access to the site at present, with the B741 to the west and a minor road lying to the east **(1 mark)**. **Possible disadvantages of this location** Large area of marsh/poorly drained land close to the site would present major construction difficulties and could result in expensive remediation work **(1 mark)**. There could be a serious threat of flooding from the loch overflowing after heavy precipitation **(1 mark)**. **Positive impacts** Creation of jobs during construction and in running the centre would be welcome in East Ayrshire where unemployment is about 1–2% above the national average, as shown in Table Q12 **(1 mark)**.

Question	General marking principle for this type of question	Max mark	Specific Marking Instruction for this question
12.	(continued)		This could boost the local economy by attracting more visitors because East Ayrshire at 18% has the lowest income from tourism in Ayrshire and Arran, as shown in Diagram Q12 (b) **(1 mark)**. This could help to reverse rural depopulation in the area, as Diagram Q12 (a) shows that there was a decrease of about 1,000 people in Dalmellington and Burnton between 2001 and 2010 **(1 mark)**.
			Negative impacts Construction work and the various activities could be very damaging to the fragile environment and could face strong opposition from conservationists **(1 mark)**. The centre and associated infrastructure could spoil the views as it is on flat land, very visible from the road **(1 mark)** and there may be an increased air and noise pollution traffic to the area harming the fragile ecosystems in the SSSI **(1 mark)**. Increased litter could lead to farm animals being harmed which could mean a financial loss for nearby farmers — Craigengillan Home Farm 475038 **(1 mark)**.

Acknowledgements

Permission has been sought from all relevant copyright holders and Hodder Gibson is grateful for the use of the following:

Image © Peter R Foster IDMA/Shutterstock.com (SQP page 14);
Image © Oneworld-images/Fotolia (Model Paper 1 page 2);
Image © Nissan South Africa (Pty) Ltd (Model Paper 1 page 6);
Photo by Simon Ledingham taken from https://commons.wikimedia.org/wiki/Sellafield#/media/File:Aerial_view_Sellafield,_Cumbria_-_geograph.org.uk_-_50827.jpg (CC BY-SA 2.0) http://creativecommons.org/licenses/by-sa/2.0/ (Model Paper 1 page 13);
An extract from the article 'Record number of radioactive particles found on beaches near Sellafield' by Rob Edwards, taken from www.theguardian.com 4 July 2012. Copyright Guardian News & Media Ltd 2015 (Model Paper 1 page 14);
Photo by Ashley Coates taken from https://www.flickr.com/photos/ashleycoates/8022935659/in/photolist-4FChFy-qwoWpB-MvZn5-ddXDfT (CC BY-SA 2.0) (Model Paper 1 page 14);
Image © iStock/Thinkstock (Model Paper 1 page 14);
Image © Sao Paulo Municipal Housing Secretary/Cities Alliance (Model Paper 2 page 3);
Population Pyramids for Ghana 2013 and 2050, taken from the US Census Bureau. Public domain (2015 page 5);
Congestion levels in UK Cities, as identified by TomTom © TomTom International B.V. (2015 page 6);
Image © gary yin/Shutterstock.com (2015 page 7);
The table 'Development indicators for selected developing countries' taken from the Central Intelligence Agency website (https://www.cia.gov/library/publications/the-world-factbook/index.html). Public domain (2015 page 10);
Diagrams taken from National Park Service U.S. Department of the Interior (http://www.nps.gov/goga/learn/nature/images/Greenhouse-effect.jpg). Public domain (2015 page 11);
Ordnance Survey maps © Crown Copyright 2015. Ordnance Survey 100047450.

Hodder Gibson would like to thank SQA for use of any past exam questions that may have been used in model papers, whether amended or in original form.